D0875887

DATE DUE

ORGANIC CHEMISTRY SERIES

Series Editor: J E Baldwin, FRS

Volume 5

Radicals in Organic Synthesis: Formation of Carbon–Carbon Bonds

Related Pergamon Titles of Interest

BOOKS

Organic Chemistry Series

DESLONGCHAMPS: Stereoelectronic Effects in Organic Chemistry
DAVIES: Organotransition Metal Chemistry: Applications to Organic Synthesis
HANESSIAN: Total Synthesis of Natural Products: the 'Chiron' Approach
PAULMIER: Selenium Reagents and Intermediates in Organic Synthesis
DEROME: Modern NMR Techniques for Chemistry Research
CARRUTHERS: Cycloaddition Reactions in Organic Synthesis

Major Works

BARTON & OLLIS: Comprehensive Organic Chemistry
KATRITZKY & REES: Comprehensive Heterocyclic Chemistry
MOO-YOUNG: Comprehensive Biotechnology
WILKINSON *et al:* Comprehensive Organometallic Chemistry

Also of Interest

BRITTON & GOODWIN: Carotenoid Chemistry and Biochemistry
COETZEE: Recommended Methods for Purification of Solvents and Tests for Impurities
HERAS & VEGA: Medicinal Chemistry Advances
KATRITZKY: Handbook of Heterocyclic Chemistry
MIYAMOTO & KEARNEY: Pesticide Chemistry: Human Welfare and the Environment
NOZAKI: Current Trends in Organic Synthesis
PERRIN *et al:* Purification of Laboratory Chemicals, 2nd Edition
RIGAUDY & KLESNEY: Nomenclature of Organic Chemistry. 'The Blue Book'
SHEMILT: Chemistry and World Food Supplies: the New Frontiers

JOURNALS

Tetrahedron (primary research journal for organic chemists)
Tetrahedron Letters (rapid publication preliminary communication journal for organic chemists)

Full details of all Pergamon publications/free specimen copy of any Pergamon journal available on request from your nearest Pergamon office.

Radicals in Organic Synthesis: Formation of Carbon–Carbon Bonds

BERND GIESE

Institut für Organische Chemie und Biochemie, Technische Hochschule Darmstadt,
Federal Republic of Germany

PERGAMON PRESS

OXFORD · NEW YORK · BEIJING · FRANKFURT
SÃO PAULO · SYDNEY · TOKYO · TORONTO

U.K.	Pergamon Press, Headington Hill Hall, Oxford OX3 0BW, England
U.S.A.	Pergamon Press, Maxwell House, Fairview Park, Elmsford, New York 10523, U.S.A.
PEOPLE'S REPUBLIC OF CHINA	Pergamon Press, Qianmen Hotel, Beijing, People's Republic of China
FEDERAL REPUBLIC OF GERMANY	Pergamon Press, Hammerweg 6, D-6242 Kronberg, Federal Republic of Germany
BRAZIL	Pergamon Editora, Rua Eça de Queiros, 346, CEP 04011, São Paulo, Brazil
AUSTRALIA	Pergamon Press Australia, P.O. Box 544, Potts Point, N.S.W. 2011, Australia
JAPAN	Pergamon Press, 8th Floor, Matsuoka Central Building, 1-7-1 Nishishinjuku, Shinjuku-ku, Tokyo 160, Japan
CANADA	Pergamon Press Canada, Suite 104, 150 Consumers Road, Willowdale, Ontario M2J 1P9, Canada

First edition 1986

Library of Congress Cataloging in Publication Data

Giese, Bernd, 1940-
Radicals in organic synthesis.
(Organic chemistry series ; v. 5)
Includes bibliographies and index.
1. Chemistry, Organic — Synthesis. 2. Radicals (Chemistry) I. Title.
II. Series: Organic chemistry series (Pergamon Press) ; v. 5.
QD262.G48 1986 547'.2 86–17052

British Library Cataloguing in Publication Data

Giese, Bernd
Radicals in organic synthesis : formation of carbon–carbon bonds. —
(Organic chemistry series ; v. 5)
1. Free radicals (Chemistry) 2. Reactivity (Chemistry) 3. Chemistry, physical organic
I. Title II. Series
547.1'224 QD471
ISBN 0–08–032493–2 Hardcover
ISBN 0–08–032494–0 Flexicover

In order to make this volume available as economically and as rapidly as possible the author's typescript has been reproduced in its original form. This method unfortunately has its typographical limitations but it is hoped that they in no way distract the reader.

Printed in Great Britain by A. Wheaton & Co. Ltd., Exeter

Foreword

The past twenty years have witnessed an unparalleled
development of new synthetic methods in the field of organic
chemistry. Many of these new methodologies involve the same
basic ionic processes which were involved in the early
development of the mechanistic picture of organic chemistry.
Although of great power these ionic processes do suffer from
limitations, in particular the problems of compatability of
functional groups. This is especially so in highly
functionalised molecules.

Within the last decade however a new approach to bond
formation has been emerging, namely the use of homolytic or
radical reactions. It is already evident that these
processes, well known of course in the polymer industry, have
a great role to play in the synthesis of complex molecules.
Professor Bernd Giese is a pioneer of this new development.
In this book he describes in a masterly fashion the
developments which have occurred and in particular how the
practising synthetic chemist can make use of them in his
work. The book will be of wide use to all chemists involved
in synthesis both in industry and academia.

J E Baldwin, FRS
Dyson Perrins Laboratory
University of Oxford

Acknowledgements

First of all I would like to thank my American postdoctorate collaborator Dr. D.B. Gerth for helping me to make my English more readable and for the numerous suggestions he has made throughout the writing of the manuscript. Furthermore, I am indebted to a large group of coworkers who have proofread the original manuscript as well as the references, namely: D. Bartmann, J.A. González-Gómez, K. Gröninger, T. Haßkerl, K. Jäger, M. Leising, W. Mehl, M. Nix, B. Rückert, R. Rupaner, G. Thoma, and T. Witzel. The manuscript was patiently typed by Mrs. E. Schmieg and the extensive art work was superbly done by Mrs. H. Roth and Mrs. E. John. Final proofreading was done by A. Ghosez and Dr. D.B. Gerth. I would also like to thank all of the colleagues who have send me preprints of their work, which have allowed me to cover literature references up to mid 1986.

Contents

CHAPTER THREE

INTERMOLECULAR FORMATION OF ALIPHATIC C-C BONDS

CHAPTER SIX

METHODS OF RADICAL FORMATION 267

Abbreviations

Ac	acetyl
AIBN	azoisobutyronitrile
Ad	adamantyl
Ar	aryl
Bn	benzyl
Bu	butyl
Bz	benzoyl
Cbz	carbobenzyloxy
Co(dmgH)$_2$py	cobaloxime (see p. 102)
DBU	diazabicycloundecane
DEAD	diethylazodicarboxylate
DHP	dihydropyran
DME	dimethoxyethane
DTBP	di-*t*-butylperoxide
e$^-$	electron
Et	ethyl
HMPA	hexamethylphosphortriamide
Im	imidazolyl
LDA	lithium diisopropylamide
MCPBA	m-chloroperbenzoic acid
Me	methyl
NBS	N-bromosuccinimide
OGlu(OAc)$_4$	tetraacetylglucoside
PCC	pyridinium chlorochromate
Pr	propyl
TBHP	*t*-butylhydroperoxide
Tf	trifluoromethanesulfonyl
THF	tetrahydrofuran
THP	tetrahydropyranyl
TMS	trimethylsilyl
Tol	p-tolyl

Chapter 1

Introduction

Radical chemistry dates back to 1900 when Gomberg[1] investi-
gated the formation and reactions of the triphenylmethyl
radical. In the 1920's Paneth[2] showed that less stabilized
alkyl radicals also exist and measured the lifetime of these
radicals in the gas phase. Organic synthesis with radicals
began in 1937 when Hey and Waters[3] described the phenylation
of aromatic compounds by benzoyl peroxide as a radical reaction.
The same year, Kharasch[4] recognized that the anti-Markovnikov
addition of hydrogen bromide to alkenes proceeds via a radical
chain process. In the following years, Mayo, Walling, and
Lewis[5] discovered the rules of radical copolymerization reactions. The results of these early investigations were
presented in two texts on radical chemistry.[6]

The deeper insights into the formation, structure, and
reactions of radicals gained in the 1950's and 60's were col-
lected (1973) in a two-volume work edited by Kochi.[7] In the
following years, the work of Ingold and others made available
the absolute rate constants of the major radical reactions in
solution. These rate data have only recently been compiled
in the Landoldt-Börnstein edited by Fischer.[8]

However, the 1970's also witnessed the start of new synthetic
methods involving radicals, particularly in substitution

reactions of aromatic compounds.[9,10] The last years have
brought a rapid development in the use of alkyl radicals for
the formation of aliphatic C-C bonds and in the synthesis of
target molecules.[11]

This monograph is an attempt to bring together the principles
that have to be followed when radicals are used in synthesis
and to demonstrate how carbon-carbon bonds can be formed in
radical reactions. Since radical chemistry provides mild
reaction conditions for the formation of C-C bonds and these
bonds constitute the backbone of organic compounds, this book
will focus on this topic.

It is suggested that the chapter on the basic principles be
read first, because one must have at least some knowledge of
the reactivity of radicals in order to successfully apply
radical reactions to synthesis.

REFERENCES

1. M. Gomberg, *J. Am. Chem. Soc.* **1900**, 22, 757; M. Gomberg
    Chem. Ber. **1900**, 33, 3150.

2. F. Paneth, W. Hofeditz, *Chem. Ber.* **1929**, 62, 1335.

3. D.H. Hey, W.A. Waters, *Chem. Rev.* **1937**, 21, 169.

4. M.S. Kharasch, E.T. Margolis, F.R. Mayo, *J. Org. Chem.*
    1937, 2, 393.

5. F.R. Mayo, F.M. Lewis, *J. Am. Chem. Soc.* **1944**, 66, 1594;
    F.R. Mayo, F.M. Lewis, C. Walling, *Discuss. Faraday Soc.*
    1947, 2, 285.

6. W.A. Waters: *The Chemistry of Free Radicals*, Clarendon
    Press, Oxford 1946; C. Walling: *Free Radicals in Solution*,
    Wiley, New York 1957.

7. J.K. Kochi (ed.): *Free Radicals*, Wiley, New York 1973.

8. H. Fischer (ed.), Landoldt-Börnstein, New Series, Vol. 13,
    Springer, Berlin since 1983.

9. J.F. Bunnett, *Acc. Chem. Res.* **1978**, 11, 413.

10. F. Minisci, *Top. Curr. Chem.* **1976**, 62, 1.

11. D.H.R. Barton, W.B. Motherwell in B.M. Trost;
C.R. Hutchinson: *Organic Synthesis Today and Tomorrow*,
Pergamon Press, Oxford 1981; D.J. Hart, *Science* **1984**,
223, 883; B. Giese, *Angew. Chem. Int. Ed. Engl.* **1985**, 24,
553; B. Giese (ed.): *Selectivity and Synthetic Applications of Radical Reactions*, Tetrahedron "Symposia-in-
Print" Number 22, *Tetrahedron* **1985**, 41, 3887 ff.

Chapter 2

Basic Principles

A. General Aspects of Syntheses with Radicals

Radicals are species with at least one unpaired electron which, in contrast to organic anions or cations, react easily with themselves in bond forming reactions. In the liquid phase most of these reactions occur with diffusion controlled rates. Radical-radical reactions can be slowed down only if radicals are stabilized by electronic effects (stable radicals) or shielded by steric effects (persistent radicals). But these effects are not strong enough to prevent diffusion controlled recombination of, for example, benzyl radicals **1**[1] or tert-butyl radicals **2**.[2] Only in extreme cases, e.g. the triphenylmethyl radical **3** or the di-tert-butyl methyl radical **4**,[3] recombination rates are low. While the recombination rate of the triphenylmethyl radical **3** is reduced due to both steric and radical stabilizing effects, the steric effect alone slows down the recombination of the di-tert-butyl methyl radical **4**. Since **3** and **4** have no C-H bonds ß to the radical center, disproportionation reactions, in which the hydrogen atom is transferred, cannot occur.

$\langle\hspace{-3pt}\overline{}\hspace{-3pt}\rangle\!-\!\dot{C}H_2$ $(CH_3)_3C\cdot$ $(C_6H_5)_3C\cdot$ $\left[(CH_3)_3C\right]_2\dot{C}H$

1   **2**   **3**   **4**

1. Reactions between radicals

The fact that reactions between radicals are in most cases
very fast could lead to the conclusion that direct radical-
radical combination is the most synthetically useful reaction
mode. This, however, is not the case because direct radical-
radical reactions have several disadvantages:

- In the recombination reactions, the radical character is
  destroyed so that one has to work with at least equivalent
  amounts of radical initiators.

- The diffusion controlled rates in radical-radical reactions
  give rise to low selectivities which cannot be influenced
  by reaction conditions.

- The concentrations of radicals are so low that reactions
  with non-radicals, like the solvents, which are present
  in high concentrations, are often hard to prevent.

Nevertheless, there are useful synthetic applications in
which new bonds are formed from radical-radical combination,
for example, the Kolbe electrolysis of carboxylates (5→7),
with the modern developments by Schäfer,[4] and the radical
induced dehydrodimerization (8→7), which has been studied
extensively in the last years by Viehe.[5]

$$R\,CO_2^- \xrightarrow[-\,CO_2]{-\,e^-} R\cdot \longrightarrow R-R$$

$$\textbf{5} \qquad\qquad\qquad \textbf{6} \qquad\qquad \textbf{7}$$

$$R-H \xrightarrow{(CH_3)_3CO\cdot} R\cdot \longrightarrow R-R$$

$$\textbf{8} \qquad\qquad\qquad \textbf{6} \qquad\qquad \textbf{7}$$

2. Reactions between radicals and non-radicals

The second method for the synthesis of products using radical chemistry employs reactions between radicals and non-radicals. It possesses the following advantages:

- The radical character is not destroyed during the reaction; therefore, one can work with catalytic amounts of radical initiators.

- Most of the reactions are not diffusion controlled, and the selectivities can be influenced by variation of the substituents.

- The concentration of the non-radicals can be easily controlled.

In most cases, in order to apply reactions between radicals and non-radicals for syntheses, chain reactions have to be built up. For the successful use of radical chains two conditions must be obeyed:

- The selectivities of the radicals involved in the chain have to differ from each other.

- The reactions between radicals and non-radicals must be faster than radical combination reactions.

These rules can best be illustrated by a chain reaction that
has gained increasing synthetic importance in the last
years.[6] In this chain reaction, alkylhalides **9** and alkenes **10**
react in the presence of tributyltin hydride to give products
12.

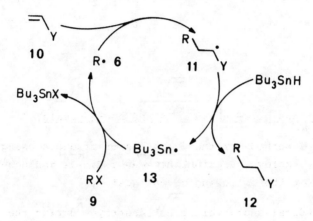

For a successful application of the tin-method, alkyl radicals
6 must attack alkenes **10** to form adduct radicals **11**. Trapping
of **11** yields products **12** and tributyltin radicals **13**, which
react with alkylhalides **9** to give back educt radicals **6**. The
tin method can be synthetically useful only if these reactions
are faster than all other possible reactions of radicals **6**,
11, and **13**. Therefore, the radicals in the chain must meet
certain selectivity and reactivity prerequisites.

a. Selectivity requirement

Radicals **6**, **11**, and **13** are simultaneously present during the
formation of product **12** from alkylhalides **9**, alkene **10**, and
tributyltin hydride. The radicals face the same competition
systems, but for a successful synthesis, each radical must

react with a specific educt. Radical **6** must add to an alkene,
adduct radical **11** has to abstract a hydrogen atom from
tributyltin hydride, and tin radical **13** must react with an
alkyl halide. This means, that the radicals in a synthetically
useful chain reaction have to differ in selectivity. If, for
example, adduct radicals **11** would possess the same selectivity
as radicals **6** ($k_1/k_2 = k_3/k_4$), then either polymerization of
alkenes ($k_1/k_2 \gg 1$), reduction of alkylhalides ($k_1/k_2 \ll 1$),
or formation of a product mixture ($k_1/k_2 \approx 1$) would result.

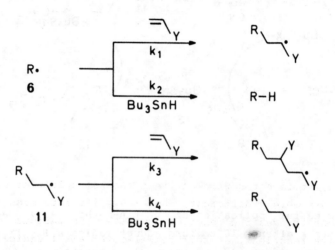

This can be prevented by choosing a suitable substituent on
the alkene that changes the selectivity of radical **11**
compared to radical **6**. Therefore, in planning syntheses one
has to know at least something about the substituent influence
on the selectivity of radicals. Fortunately, the mechanism
of radical addition to alkenes is known in detail.[7] Alkyl
radicals, substituted with electron-releasing groups (alkyl,
alkoxy, amino etc.), behave like nucleophiles and react very
fast with alkenes substituted by electron-withdrawing
substituents (nitrile, ketone, ester etc.).[7,8] On the other

hand, radicals with electron-withdrawing substituents behave
like electrophiles and react fast with electron-rich
alkenes.[7,9] The synthetic application of the tin-method is
possible, therefore, only with those halides **9** and alkenes **10**
which lead to radicals **6** and **11** with different or opposite
polarity. For example, one can use the pairs cyclohexyl
iodide/acrylonitrile[10] or chloromalonic ester/enolether.[11]

Of course, it is not sufficient that the selectivities of the
radicals are merely different; they must differ to such a
degree that the desired reaction dominates the various
competitive reactions. At room temperature, alkyl radicals R·
add to alkenes such as acrylonitrile with rate coefficients of
about 10^5 - 10^6 (1/mol·s),[8] whereas nitrile-substituted
radicals react with acrylonitrile about 10^4 times slower.[12]
Hydrogen transfer from Bu_3SnH to alkyl radicals also occurs
with rate coefficients of 10^5 - 10^6 (1/mol·s) at room temperature,[13] but these rates don't depend very much on the
substituent at the radical center.[14] Therefore, the
selectivities of alkyl radicals **6** and adduct radicals **11** in
the competition system $H_2C=CHCN/Bu_3SnH$ are k_1/k_2 = 1 and
k_3/k_4 = 10^{-4}. In order to form a C-C bond, the reaction between
radical **6** and the alkene has to be faster than the reduction
of **6** by Bu_3SnH. This is possible only with a $H_2C=CHCN/Bu_3SnH$

concentration ratio of at least 10-100, because the
competition constant is about 1. Therefore, either one must
work with an excess of acrylonitrile, or with low concentrations of Bu_3SnH. Low Bu_3SnH concentrations can be attained
by slowly adding the tin hydride to the reaction mixture or
by working with a catalytic amount of tin salts and equimolar
amounts of $NaBH_4$, which generates the tin hydride in situ.
When a $H_2C=CHCN/Bu_3SnH$ ratio of 10^2 is used, polymerization
doesn't occur because the competition coefficient k_4/k_3 is
about 10^4. The rate ratio r_4/r_3,which takes the concentrations
into account, is 10^2; therefore, adduct radical **11** reacts
predominately with Bu_3SnH to yield product **12**. If the
excess of acrylonitrile is too large, then **11** reacts noticeably
with acrylonitrile. An essential task, therefore, is to find
the optimum concentration ratio for the synthesis in question.
This is necessary because alkene substituents exert a strong
influence on the rate. Thus, the rate of addition of the
cyclohexyl radical increases by 8500 on going from 1-hexene
to acrolein.[7,15]

The third radical involved in the chain is the tributyltin
radical **13**. For this radical the selectivity in the
competition system alkene/tributyltin hydride is of no
importance, because the reaction of **13** with tin hydride
generates back the tin radical. However, the competition
between alkene addition and halogen abstraction from the alkyl
halide is synthetically important.

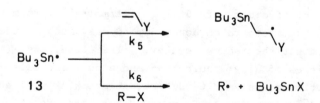

In fact, alkyl bromides and alkenes like acrylonitrile, react equally fast with the tin radical **13**.[16] Since a relatively large amount of acrylonitrile is already used in these syntheses, the undesired addition of tin radicals **13** to alkenes cannot be greatly influenced by varying the concentration ratio alkene/alkyl halide. Instead, it is better to switch from alkylbromides to the 10-100 times more reactive alkyl iodides.[16]

C-C bond formation doesn't take place when less reactive alkyl halides are used because alkyl radicals **6** are not regenerated in the chain. Therefore, not only alkyl chlorides, but also isonitriles that are reduced by tin hydride cannot be used as educts.[17] Syntheses with these educts in the presence of alkenes lead to tin hydride addition products on the alkenes.

$$RCl \; + \; Bu_3SnH \; \longrightarrow \; RH \; + \; Bu_3SnCl$$

This analysis is typical for the development of radical chain reactions in syntheses. In order to evaluate which reactions are synthetically useful, the magnitudes of the rate constants must be known. Many of these rate constants have been determined in recent years and most of the important ones are now known. Moreover, the solvent has little effect on the reaction rates of neutral radicals; therefore, the literature data can be used even if the synthesis is carried out in a different solvent.

b) Reactivity requirement

Chain reactions are terminated by combination of the radicals.
Therefore, the rate of chain propagation between radicals and
non-radicals (r_p) must be higher than that of chain termination
between the radicals (r_t). Equations (a) and (b) are the
analytical expressions for these reactions, where k_p and k_t
are the rate constants for chain propagation and chain
termination steps, respectively. R· represents all radicals
that are present in the chain, and X the added reagents. The
reactivity requirement that must be satisfied for the use of
radical chains in syntheses is given by equation (c).

$$r_p = k_p \cdot [R\cdot] \cdot [X] \qquad (a)$$

$$r_t = k_t \cdot [R\cdot] \cdot [R\cdot] \qquad (b)$$

$$1 < \frac{r_p}{r_t} = \frac{k_p \cdot [X]}{k_t \cdot [R\cdot]} \qquad (c)$$

In most cases, carbon centered radicals react with each other
with diffusion-controlled rates; the rate coefficients k_t,
which have only a small temperature dependence, are $10^9 - 10^{10}$
(1/mol·s) in the liquid phase.[1,2] The concentration of the
added reaction partner X depends on the reaction conditions,
but in syntheses, these concentrations are often nearly one
molar. The concentrations of radicals in chain reactions also
depend on the reaction conditions, e.g. the rate of the
decomposition of the initiator. Typically, radical concentrations in chain reactions are about $10^{-7} - 10^{-8}$ (mol/l).[18]

$$k_p > 10^2 \text{ (1/mol·s)} \qquad (d)$$

According to expression (d), only those chain propagation
steps whose rate coefficients k_p are larger than 10^2 (1/mol·s)
can be used for syntheses. Thus, the homolytic cleavage of
O-H and N-H bonds in aliphatic alcohols and amines by alkyl

radicals, as well as the intermolecular addition to the C=O group of ketones and esters are so slow at room temperature that these reactions don't occur in synthetically useful chains. In other words, carbon-centered radicals exhibit high chemoselectivities and their employment in reactions with more complex molecules should be considered.

B. Elementary Reaction Steps between Radicals and Non-Radicals

2. Introduction

A radical chain is built up by different types of propagation steps all of which lead to new radicals

- Addition reactions.

$$R\cdot\ +\ AB\ \rightarrow\ RAB\cdot$$

- Substitution (abstraction) reactions.

$$R\cdot\ +\ AB\ \rightarrow\ RA\ +\ B\cdot$$

- Elimination (fragmentation) reactions.

$$RAB\cdot\ \rightarrow\ RA\ +\ B\cdot$$

- Rearrangement reactions.

$$RAB\cdot\ \rightarrow\ ARB\cdot$$

- Electron transfer reactions.

$$R^-\ +\ M^{n+}\ \rightarrow\ R\cdot\ +\ M^{(n-1)+}$$

or

$$RX^{\cdot-}\ +\ RY\ \rightarrow\ RX\ +\ RY^{\cdot-}$$

It is not the goal of this book to go into the mechanistic
details of these elementary steps, but very often two simple
rules can be applied to fast radical chain reactions:

- Most chain propagating steps are exothermic and one can
  use the strength of bonds that are broken and formed as a
  rough guideline for the rate of the reaction (thermodynamic
  parameter).

- Because of the early transition states in fast radical
  reactions, frontier molecular orbital (FMO) theory can be
  utilized for these reactions (kinetic parameter).

The applicability of these rules shall be demonstrated by some
examples for the five different propagation steps.

2. Addition

Addition of alkyl radicals to alkenes is a useful C-C bond
formation reaction in which a σ-CC bond is made from a
π-CC bond in a very exothermic reaction. In contrast, π-CO
bonds of ketones and aldehydes are nearly as strong as σ-CC
bonds. Therefore, ketones and aldehydes cannot be used as
intermolecular traps in syntheses.

$$R\cdot \quad + \quad \underset{/}{\overset{\backslash}{C}}=\underset{\underset{Y}{\backslash}}{\overset{/}{C}} \quad \xrightarrow{\text{useful}} \quad R-\overset{|}{\underset{|}{C}}-\underset{\backslash Y}{\overset{/}{C}}\cdot$$

$$R\cdot \quad + \quad \underset{/}{\overset{\backslash}{C}}=O \quad \xrightarrow{\text{not useful}} \quad R-\overset{|}{\underset{|}{C}}-O\cdot$$

The rate of addition of a radical to an alkene depends largely on the substituents on the radical and the alkene. These substituent effects can be described by FMO theory.[7,19] The singly occupied orbital (SOMO) of the radical interacts with the lowest unoccupied orbital (LUMO) and/or the highest occupied orbital (HOMO) of the CC-multiple bond. Radicals with a high lying SOMO interact preferentially with the LUMO of the alkene.

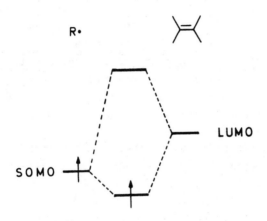

Orbital interaction between a nucleophilic
radical and an electron-poor alkene

Electron withdrawing substituents at the alkene, which lower the LUMO energy, increase the addition rate by reducing the SOMO-LUMO difference. Therefore, cyclohexyl radicals react 8500 times faster with acrolein than with 1-hexene.[15]

$$\underset{=}{\diagup}C_4H_9 \qquad \underset{=}{\diagup}C_6H_5 \qquad \underset{=}{\diagup}CO_2CH_3 \qquad \underset{=}{\diagup}CHO$$

$$k \,\overset{\bullet}{C_6}H_{11} \equiv 1.0 \qquad\qquad 84 \qquad\qquad 3000 \qquad\qquad 8500$$

These orbital effects are so important that a tert-butyl radical reacts faster than prim. and sec. radicals with alkenes like vinylphosphonic ester or acrylonitrile.[20] Thus, the increase in the SOMO energy in going from prim. to tert. radicals has a larger effect on the rate than the decrease in the strength of the bonds that are formed. Alkyl, alkoxyalkyl, aminoalkyl, and other similar radicals are therefore nucleo- philes. However, radicals with electron withdrawing substituents at the radical center have SOMO energies so low that the SOMO-HOMO interaction dominates.

Orbital interaction between an electrophilic radical and an electron-rich alkene

These radicals react like electrophiles; that is, electron
donating substituents at the alkenes increase the rate. The
malonyl radical, for example, reacts with enamine **14** 23 times
faster than with acrylester **15**.[11]

$$
\begin{array}{ccc}
\underset{\textstyle\mathbf{14}}{\chemfig{C_6H_5 \cdots N \cdots O}} & \underset{\textstyle\quad}{\chemfig{C_6H_5 \cdots CH_3}} & \underset{\textstyle\mathbf{15}}{\chemfig{C_6H_5 \cdots CO_2C_2H_5}}
\end{array}
$$

$$^k(EtO_2C)_2\overset{\bullet}{C}H \qquad 23 \qquad\qquad 3.5 \qquad\qquad \equiv 1.0$$

3. Substitution

Since an O-H bond is much stronger than a C-H bond, a typical
bimolecular reaction for alkoxy radicals is the H-abstraction
from C-H bonds. However, OH-bonds of alcohols are attacked
too slowly for synthetic applications because the reaction is
thermoneutral.

$$RO\cdot \; + \; -\!\overset{\textstyle |}{\underset{\textstyle |}{C}}\!-H \quad\xrightarrow{\text{useful}}\quad ROH \; + \; -\!\overset{\textstyle |}{\underset{\textstyle |}{C}}\!\cdot$$

$$RO\cdot \; + \; -\!\overset{\textstyle |}{\underset{\textstyle |}{C}}\!-OH \quad\xrightarrow{\text{not useful}}\quad ROH \; + \; -\!\overset{\textstyle |}{\underset{\textstyle |}{C}}\!-O\cdot$$

Alkoxy radicals are electrophiles and they preferentially
attack C-H bonds with high HOMO energies; for instance, the
α-C-H bond of ethers and amines or the alkyl C-H bond of
esters.[21] In contrast, nucleophilic alkyl radicals abstract
a hydrogen atom from the acyl group of esters, because this
C-H bond has a lower LUMO energy.

$$R\overset{\bullet}{C}HCO_2CH_2R \quad \overset{R\cdot}{\longleftarrow} \quad RCH_2\overset{O}{\overset{\|}{C}}-OCH_2R \quad \overset{RO\cdot}{\longrightarrow} \quad RCH_2CO_2\overset{\bullet}{C}HR$$

$$CH_3\overset{\bullet}{C}HCO_2H \quad \overset{\overset{\bullet}{C}H_3}{\longleftarrow} \quad CH_3CH_2CO_2H \quad \overset{Cl\cdot}{\longrightarrow} \quad \overset{\bullet}{C}H_2CH_2CO_2H$$

These differences also account for the preferential abstraction
of the ß-hydrogen of propionic acid by the electrophilic
chlorine atom and the abstraction of the α-hydrogen by the
nucleophilic methyl radical.[21]

3. Elimination

In elimination reactions, two molecules are formed from one.
Thus, these reactions are favored by activation entropies, and
the free energy gain increases with increasing reaction
temperature. Therefore, even alkoxy radicals undergo fast
ß-elimination reactions, although the enthalpy differences
between π-CO and σ-CC bonds are small.[22] But a C-OR bond α
to a radical center is cleaved too slowly to be of synthetic
use, because less stable π-CC bonds are formed. Only with the
weaker C-Br, C-SR or C-SnR$_3$ bonds are ß-elimination reactions
fast enough to be synthetically useful.[23]

$$R-\overset{|}{\underset{|}{C}}-O\cdot \quad \overset{\text{useful}}{\longrightarrow} \quad R\cdot \;+\; {>}C{=}O$$

$$\overset{RO}{\underset{|}{\overset{|}{-C}}}-C\overset{\bullet}{<} \quad \overset{\text{not useful}}{\longrightarrow} \quad RO\cdot \;+\; {>}C{=}C{<}$$

5. Rearrangement

Compared to their cationic counterparts, only a few radical
rearrangements are fast enough to appear in syntheses.[24] An
example is the vinyl migration in **16** which forms the more
stable radical **18**.

| 16 | 17 | 18 |

This reaction is a combination of an intramolecular addition
(**16→17**) and an elimination reaction (**17→18**). The addition
of the nucleophilic radical to the electron-rich alkene is
fast because the loss in entropy is much smaller than in
intermolecular reactions. The ß-C-C bond of **17** cleaves rapidly
because of ring strain. The much smaller ring strain in **20**
stops the rearrangement of the hexenyl radical **19** at the
cyclopentylmethyl radical stage.

| 19 | | 20 |

This is in accord with the very slow 1,2 alkyl or hydrogen
shifts in radicals, which don't occur in syntheses because
of the strong σ-bonds. An orbital effect presumably accounts
for the formyl shift (**21→22**), which is nearly as fast as the
vinyl shift of **16**,[25] even though a π-CO bond is much stronger
than a π-CC bond.

21   **22**

6. Electron transfer

The rate of electron transfer reactions depends on the
difference in the reduction potentials of educts and products.
Since alkyl radicals possess an unpaired electron in a
nonbonding orbital, electron transfer reactions to many metal
salts often occur with high rates.[26] The higher the SOMO
energies of the radicals, the faster is the electron transfer.
The alternating addition of alkyl radicals to olefin pairs,
observed by Minisci, is a significant example of this
behavior.[27] Alkyl radicals generated by electron transfer
from Cu^+ to peroxides form adduct radicals **23** with acrylonitrile. The nitrile substituent at the radical center of **23**
causes oxidation to be slower than addition to styrene and
gives the 2:1 adduct radical **24**. Electron transfer from
this benzylic radical is now faster than further addition
to the alkene and products are formed via electron transfer
and reaction with alcohols.

peroxide   + Cu^+ $\longrightarrow$ R• + Cu^{2+}

R•   + $\diagdown\!\!\diagup^{CN}$ $\longrightarrow$ $R\diagup\!\!\diagdown\diagup\overset{\bullet}{_{CN}}$

23

$R\diagdown\diagup\overset{\bullet}{_{CN}}$ + $\diagdown\!\!\diagup^{C_6H_5}$ $\longrightarrow$ $R\diagup\!\!\diagdown\underset{CN}{\diagup}\diagdown\overset{\bullet}{\diagup^{C_6H_5}}$

24

$R\diagup\!\!\diagdown\underset{CN}{\diagup}\diagdown\overset{\bullet}{\diagup^{C_6H_5}}$ + Cu^{2+} $\longrightarrow$ $R\diagup\!\!\diagdown\underset{CN}{\diagup}\diagdown\overset{+}{\diagup^{C_6H_5}}$

$R\diagup\!\!\diagdown\underset{CN}{\diagup}\diagdown\overset{+}{\diagup^{C_6H_5}}$ + $R'OH$ $\longrightarrow$ $R\diagup\!\!\diagdown\underset{CN}{\diagup}\diagdown\underset{OR'}{\diagup^{C_6H_5}}$

Compared to neutral radicals, electron transfer reactions with radical anions are so fast, that even neutral molecules can act as electron acceptors. This is the case in the $S_{RN}1$ reactions developed by Kornblum, Russell and Bunnett.[28] Thus, indoles **26** can be synthesized from aniline derivatives **25** in high yields via a multistep reaction sequence involving electron transfer reactions between radical anion **27** and the starting aniline **25** (see p. 252).

25   **26**

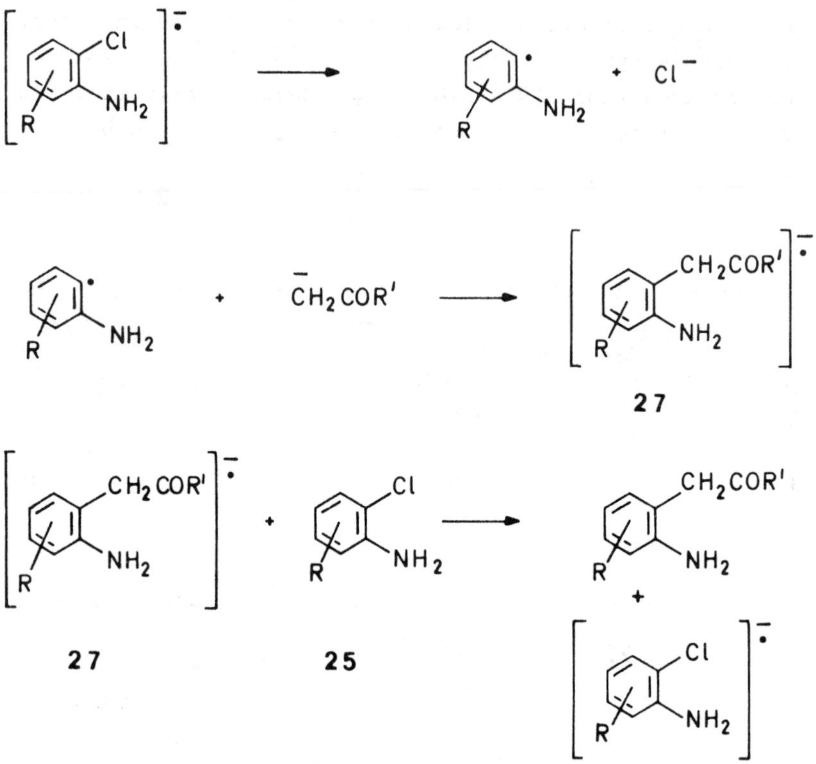

27

27   **25**

C.   Comparison of Radicals and Ions in Syntheses

1.   Chemoselectivity

Because of the reactivity requirement only functional groups
that have bimolecular rate coefficients $k > 10^2$ (1/mol·s) are
attacked by radicals. The limiting factor can be advantageous
in the planning of syntheses because it means that a range of
functional groups are tolerated without protection. Thus, the
homolytical cleavage of O-H and N-H bonds in aliphatic
alcohols and amines by alkyl radicals, as well as the

intermolecular addition of alkyl radicals to C=O bonds of
ketones and esters are so slow at room temperature that they
don't occur in synthetically useful chains. In other words,
carbon-centered radicals exhibit high chemoselectivities and
their use in reactions with complex molecules should be
considered.

28   **29**   45 %   **30**

29   Bu₃SnH

31

For example, the C-branched carbohydrate **30** can be synthesized
from the glucal **28** without protection of the OH groups. The
C-C bond formation occurs via homolysis of the C-Hg bond in
29, attack of fumaronitrile at radical **31**, and subsequent
hydrogen transfer, to yield product **30**.[29]

Another example is the cyclization of **32** to product **33**
carried out by Stork.[30] Again there is no need to protect
the OH group during the synthesis. It is also remarkable to note
how facile the quaternary carbon atom in **33** is formed. The
reaction occurs via the vinyl radical **34**, which exclusively
cyclizes to a five-membered ring **35**; hydrogen abstraction then
gives product **33**.

2. Regioselectivity

Radical reactions often show different regioselectivities
than ionic reactions. Thus, α,β-unsaturated aldehydes, ketones,
and esters are attacked by carbon centered radicals
exclusively at the olefinic carbon atom. In contrast, anionic
species exhibit a competition between attack at the olefinic
and the carbonyl C-atom.

*Attack of radicals and anions at the LUMO of α,β-unsaturated
carbonyl systems. The circles describe the quantity of the
orbital coefficients.*

Radicals attack exclusively at the carbon atom with the largest
LUMO coefficient. The resulting higher regioselectivity of
radicals in addition reactions to α,ß-unsaturated carbonyl
compounds demonstrates that orbital interactions are more
important in radical than in ionic reactions.

Another special feature of radicals is the predominant
cyclization of 1-hexenyl radicals to five-membered rings, [24,31]
whereas cations yield six-membered rings.

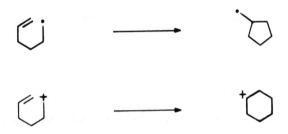

This differing behavior is caused by the different transition
states for radical and cationic reactions. For radicals,
kinetic measurements[7] and theoretical calculations[24] favor an
unsymmetrical transition state in which the distances between
the attacking radical and the two olefinic carbon atoms of
the alkene are unequal.[32] In contrast, cations attack the
center of the double bond where the electron density is
high. Beckwith has shown that, because of ring strain effects,
an unsymmetrical transition state leads to faster cyclization
to a five-membered than to a six-membered ring.[31] This
cyclization mode of radicals is also supported by the
Baldwin-rules.[33]

The formation of five-membered rings during radical cyclization
has been used extensively in the last years for the synthesis
of target molecules. Thus, Curran[34] synthesized tricyclus **37**

by homolytic iodine abstraction from **36**. The cationic
cyclization of **38** yields the six-membered rings of **39**.[35]

36   **37**   60 %

38   **39**   30 %

3. Stereoselectivity

Only a few examples of high stereoselectivities of acyclic
radicals are known. For example, radical HBr addition to
alkenes can occur in a *trans*-reaction, because of the bridging
ability of the bromine located ß to the radical center. Thus,
at -80°C the addition of HBr to *cis*- and *trans*-2-bromo-2-
butene quantitatively gives products resulting from *trans*-
addition: The *cis*-olefin **40** yields pure meso-2,3-dibromobutane
41 and the *trans*-alkene **42** gives pure d,l-dibromide **43**.[36]

By raising the temperature, the stereoselectivity decreases
so that at room temperature roughly the same mixture of
products is obtained from each olefin. In general, the
bridging of radicals like **44** is much less pronounced than that
of cations;[36,37] thus, the stereoselectivity of acyclic
radicals is often lower than that of cations. However, high
stereoselectivities can be observed with cyclic radicals. For
example, in C-C bond forming reactions, cyclopentyl radicals
45 give exclusively *trans*-adducts if the neighboring
substituent is an amide group.[38] With less shielding CH_3 or
OR groups, the stereoselectivity decreases, but it raises again
if alkenes of low reactivity or with substituents at the
attacked olefinic C-atom are used.[39] With cyclohexyl radicals
46, the *trans*-stereoselectivity is less pronounced, but can
also be relatively high.[38,39]

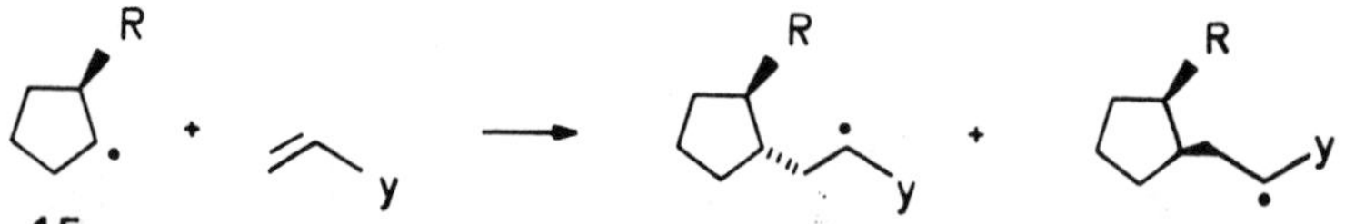

45

R	Alkene			
OCH_3	$H_2C=CHCN$	77	:	23
CH_3	$H_2C=CHCN$	92	:	8
NHAc	$H_2C=CClCN$	> 98		
OCH_3	$H_2C=CHCO_2CH_3$	88	:	12
OCH_3	$EtO_2C\diagdown\diagup CO_2Et$	98	:	2

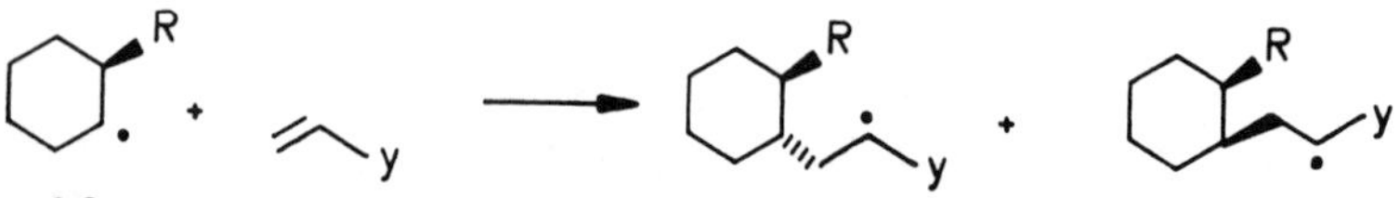

46

R	Alkene			
OCH_3	$H_2C=CHCN$	65	:	35
NHAc	$H_2C=CClCN$	> 95		
OCH_3	$H_2C=CHCO_2CH_3$	70	:	30
OCH_3	$EtO_2C\diagdown\diagup CO_2Et$	92	:	8

It is interesting to note that carbohydrate radicals, for
example **47**, show higher selectivities than cyclohexyl
radicals.[10,29]

47

85 : 15

> 98

4. Umpolung

In radical reactions the product of "umpolung" of the
reactivity can be often observed.[25,40] If one, for example,
generates ions from α-haloethers **48**, cations **49** will be formed,
because cations are stabilized by the alkoxy substituent.
These cations are, of course, electrophiles and add easily to
electron-rich alkenes like enolethers. The corresponding
alkoxy alkyl radical **50**, however, is nucleophilic, and because
of its high lying SOMO, it preferentially attacks electron-
poor alkenes like acrylonitrile. Thus, α-bromoglucose **51**
gives the 1,4-di-heterosubstituted product **52** in a radical
reaction,[41] but the 1,3-di-heterosubstituted compound **53** in
an ionic reaction.[42]

Carbanions **54**, generated from malonates are stabilized by the ester substituents and are nucleophiles that undergo Michael addition with electron poor alkenes. But the corresponding malonyl radicals **55** are electrophiles, which, because of their low lying SOMO, easily attack enolethers.[11] Again radicals yield 1,4-di-heterosubstituted products, whereas ions form 1,5-di-heterosubstituted products.

$$(RO_2C)_2\overset{\bullet}{C}H \quad + \quad \diagup\diagdown_{OR} \quad \longrightarrow \quad RO_2C\diagup\overset{CO_2R}{\diagdown}\overset{\bullet}{\diagup}\diagdown_{OR}$$

55

5. Racemization

Since radical reactions used in syntheses are very fast,[6,43]
side reactions, e.g. radical rearrangements, ß-elimination,
and racemizations of chiral centers at adjacent or remote
carbon atoms, are in most cases suppressed. Moreover, most
radical reactions are carried out in the absence of strong
acids and bases so that competing ionic reactions such as
racemization don't occur. This means, that the radical
methodology discussed here offers very mild reaction conditions
under which chiral centers at the non-radical carbon atoms
survive.

In many naturally occuring chiral molecules the chiral centers
are substituted by OR or NR_2 groups. Ionic centers generated
adjacent to an OR group can cause problems; a carbanion can
induce heterolytic cleavage of the C-O bond, and with
carbocations, rearrangement can occur.

$$-\overset{-}{\underset{|}{C}}-\overset{|}{\underset{|}{C}}-\quad\longrightarrow\quad \diagup^{C=C}\diagdown \quad + \quad RO^-$$
$$\quad\quad\underset{OR}{}$$

$$-\overset{+}{\underset{|}{C}}-\overset{|}{\underset{|}{C}}-\quad\rightleftharpoons\quad -\overset{|}{\underset{|}{C}}-\overset{+}{\underset{|}{C}}-$$
$$\quad\quad\underset{OR}{}\quad\quad\quad\quad\quad\underset{OR}{}$$

These side-reactions make intermolecular reactions at ionic
centers often impossible or lead to racemized products. In
radicals with ß-OR or ß-NR$_2$ groups, elimination or racemization
cannot compete with the fast chain reactions. This has been
utilized, for example, in syntheses where the chiral centers
are delivered from the chiral pool. Thus, from lactic acid **56**,
the lactone **57**,[44] and from amino acid **58**, the aminoadipic
acid **59**[45] can be synthesized without racemization. Radical
C-C bond formation is the main step in these syntheses.

Elimination can compete with radical chain steps only if
the ß-bonds are very weak, e.g. C-S, C-Br or C-metal
bonds.[23,46]

REFERENCES

1. M. Lehni, H.Schuh, H. Fischer, *Int. J. Chem. Kinetics*
    1979, *11*, 705; H. Langhals, H. Fischer, *Chem. Ber.* **1978**,
    111, 543.

2. H.R. Dütsch, H. Fischer, *Int. J. Chem. Kinetics* **1981**, 13,
    527.

3. D. Griller, K.U. Ingold, Acc. Chem. Res. **1976**, 9,13;
    H.D. Beckhaus, G. Hellmann, C. Rüchardt, Chem. Ber.
    1978, 111, 72; K. Schlüter, A. Berndt, Tetrahedron Lett.
    1979, 929.

4. H.J. Schäfer, Angew. Chem. Int. Ed. Engl. **1981**, 20, 911.

5. H.G. Viehe, R. Merenyi, L. Stella, Z. Janousek, Angew.
    Chem. Int. Ed. Engl. **1979**, 18, 917.

6. B. Giese, Angew. Chem. Int. Ed. Engl. **1985**, 24,553.

7. B. Giese, Angew. Chem. Int. Ed. Engl. **1983**, 22, 771.

8. A. Citterio, A. Arnoldi, F. Minisci, J. Org. Chem. **1979**,
    15, 2674; B. Giese, G. Kretzschmar, Chem. Ber. **1984**, 117,
    3160; K. Münger, H. Fischer, Int. J. Chem. Kinetics
    1985, 17, 809.

9. J.M. Tedder, J.C. Walton, Tetrahedron **1980**, 36, 701;
    K. Riemschneider, E. Drechsel-Grau, P. Boldt, Tetrahedron
    Lett. **1979**, 185.

10. B, Giese, J.A. González-Gómez, T. Witzel, Angew. Chem.
    Int. Ed. Engl. **1984**, 23, 69.

11. B. Giese, H. Horler, M. Leising, Chem. Ber. **1986**,119,444.

12. R. Korus, K.F. O'Driscoll in J. Brandrup, E.H. Immergut:
    Polymer Handbook, Wiley, New York 1975, p. II-45.

13. C. Chatgilialoglu, K.U. Ingold, J.C. Scaiano, J. Am.
    Chem. Soc. **1981**, 103, 7739.

14. H. Sakurai in J.K. Kochi: Free Radicals, Vol. 2, Wiley,
    New York 1973, p. 770.

15. B. Giese, G. Kretzschmar, Chem. Ber. **1983**, 116, 3267.

16. K.U. Ingold, J. Lusztyk, J.C. Scaiano, J. Am. Chem. Soc.
    1984, 106, 343.

17. D.H.R. Barton, G. Bringmann, G. Lamotte, R.S. Hay-Mother-
    well, W.B. Motherwell, Tetrahedron Lett. **1979**, 2291.

18. K.U. Ingold in J. Kochi: Free Radicals, Vol. 1, Wiley,
    New York 1973, p. 37.

19. I. Fleming: *Frontier Orbitals and Organic Chemical Reactions*, Wiley, London 1976.

20. T. Caronna, A. Citterio, M. Ghirardini, F. Minisci, *Tetrahedron* **1977**, *33*, 793; J.A. Baban, B.P. Roberts, *J. Chem. Soc. Perkin Trans. 2*, **1981**, 161.

21. J. Tedder, *Angew. Chem. Int. Ed. Engl.* **1982**, *21*, 401.

22. P. Brun, B. Waegell in R.A. Abramovitch: *Reactive Intermediates*, Vol. 3, Plenum Press, New York 1983, p. 392.

23. J.A. Kerr, A.C. Lloyd, *Quart. Rev. (London)*, *1968*, *22*, 549; T.E. Boothe, J.L. Greene, P.B. Shevlin, *J. Am. Chem. Soc.* **1976**, *98*, 951; G.E. Keck, E.J. Enholm, J.B. Yates, M.R. Wiley, *Tetrahedron*, **1985**, *41*, 4079.

24. A.L.J. Beckwith, K.U. Ingold in P. de Mayo: *Rearrangement in Ground and Excited States*, Vol. 1, Academic Press, New York 1980, p. 161.

25. B. Giese, H. Horler, *Tetrahedron Lett.* **1983**, *24*, 3221.

26. J.K. Kochi: *Organometallic Mechanism and Catalysis*, Academic Press, New York 1978; M. Julliard, M. Chanon, *Chem. Rev.* **1983**, *83*, 425.

27. F. Minisci, *Acc. Chem. Res.* **1975**, *8*, 165.

28. R.C. Kerber, G.W. Urry, N. Kornblum, *J. Am. Chem. Soc.* **1965**, *87*, 4520; G.A. Russell, W.C. Danen, *J. Am. Chem. Soc.* **1966**, *88*, 5663; J.K. Kim, J.F. Bunnett, *J. Am. Chem. Soc.* **1970**, *92*, 7463; R.A. Rossi, R.H. de Rossi, *ACS Monogr.* *178*, **1983**.

29. B. Giese, K. Gröninger, *Tetrahedron Lett.* **1984**, *25*, 2743.

30. G. Stork, N.H. Baine, *J. Am. Chem. Soc.* **1982**, *104*, 2321.

31. A.L.J. Beckwith, C.H. Schiesser, *Tetrahedron* **1985**, *41*, 3925.

32. H. Fujimoto, S. Yamabe, T. Minato, K. Fukui, *J. Am. Chem. Soc.* **1972**, *94*, 9205; M.J.S. Dewar, S. Olivella. *J. Am. Chem. Soc.* **1978**, *100*, 5290.

33. J.E. Baldwin, *J. Chem. Soc. Commun.* **1976**, 734.

34. D.P. Curran, D.M. Rakiewicz, *Tetrahedron* **1985**, *41*, 3943.

35. W.S. Johnson, M.F. Semmelhack, M.U.S. Sultanbawa, L.A. Dolak, *J. Am. Chem. Soc.* **1968**, *90*, 2994.

36. P.S. Skell, K.J. Shea in J.K. Kochi: *Free Radicals*, Vol. 2, New York 1973, p. 809.

37. J.W. Wilt in J.K. Kochi: *Free Radicals*, Vol. 1, New York 1973, p. 333.

38. R. Henning, H. Urbach, *Tetrahedron Lett.* **1983**, *24*, 5343.

39. B. Giese, K. Heuck, H. Lenhardt, U. Lüning, *Chem. Ber.* **1984**, *117* 2132; B. Giese, H. Harnisch, U. Lüning, *Chem. Ber.* **1985**, *118*, 1345.

40. B. Giese, H. Horler, *Tetrahedron* **1985**, *41*, 4025.

41. B. Giese, J. Dupuis, *Angew. Chem. Int. Ed. Engl.* **1983**, *22*, 622; J. Dupuis, B. Giese, D. Rüegge, H. Fischer, H.-G. Korth, R. Sustmann, *Angew. Chem. Int. Ed. Engl.* **1984**, *23*, 896.

42. R.R. Schmidt, M. Hoffmann, *Angew. Chem. Int. Ed. Engl.* **1983**, *22*, 406.

43. C. Walling, *Tetrahedron* **1985**, *41*, 3887.

44. B. Giese, D.B. Gerth, *J. Org. Chem.* **1986**, in press.

45. R.M. Adlington, J.E. Baldwin, A. Basak, R.P. Kozyrod, *J. Chem. Soc. Chem. Commun.* **1983**, 944.

46. M.D. Johnson, *Acc. Chem. Res.* **1983**, *16*, 343; A. Gaudemer, K. Nguyen-Van-Duong, N. Shahkarami, S.S. Achi, M. Frostin-Rio, D. Pujol, *Tetrahedron* **1985**, *41*, 4095.

Chapter 3

Intermolecular Formation of Aliphatic C—C Bonds

A. Introduction

The most important methodology for the synthesis of aliphatic C-C bonds via radical reactions is the addition of alkyl radicals **1** to alkenes **2**. This reaction leads to adduct radicals **3** that must be converted to non-radical products before polymerization occurs. Polymerization is avoided either by intermolecular trapping of adduct radicals **3** or by intra-molecular, homolytic bond cleavage. Hydrogen atom donors X-H, heteroatom donors X-Z or electron donors M^{n+} are used as trapping agents.

$$
R\cdot + \underset{\mathbf{2}}{\diagup C = C \diagdown_{Y}} \longrightarrow \underset{\mathbf{3}}{R - \overset{|}{\underset{|}{C}} - \overset{|}{\underset{Y}{C}}\cdot}
$$

$$
\xrightarrow[\text{-X}\cdot]{HX} \quad R - \overset{|}{\underset{|}{C}} - \overset{|}{\underset{Y}{C}} - H \quad \mathbf{4}
$$

$$
\xrightarrow[\text{-X}\cdot]{XZ} \quad R - \overset{|}{\underset{|}{C}} - \overset{|}{\underset{Y}{C}} - Z \quad \mathbf{5}
$$

$$
\xrightarrow[\text{-M}^{(n-1)+}]{M^{n+}} \quad R - \overset{|}{\underset{|}{C}} - \overset{|}{\underset{Y}{C}} + \quad \mathbf{6}
$$

with **1** at left of **2**.

In the alkene/radical trap competition system, educt radicals **1**
must react faster with alkenes, and adduct radicals **3** must
react faster with the radical traps. If this is not the case,
either educt radicals are trapped before they can form a C-C
bond or adduct radicals react with alkenes to give polymers.
This selectivity requirement can be fulfilled by choosing
suitably substituted alkenes. With nucleophilic alkyl radicals
1 one has to use alkenes **2** with electron-withdrawing groups Y
(e.g. acrylonitrile) that reduce the nucleophilicity of the
adduct radicals **3**, whereas with electrophilic radicals **1**,
alkenes with electron-releasing substituents (e.g. enolethers)
must be used. These selectivity changes reduce the amount of
polymerization because the more nucleophilic the radical is,
the faster is the reaction with an electron-poor alkene and
vice versa.[1] If the adduct radicals give products by homolytic
bond cleavage, this selectivity change is of less importance
because now an intramolecular reaction competes with intermolecular reactions.

After adduct radicals **3** or **8** are converted to products the
newly formed radicals X· must react with the radical precursor
to give back the starting radicals R·. This condition severely
limits the application of radical traps in syntheses. In some
cases the newly formed radical X· is identical with the
educt radical **1**. This simplifies the radical chain, but also
reduces the flexibility of the synthesis.

In spite of all these conditions, several methods for the formation of C-C bonds using radical chemistry have been worked out and successfully applied to the synthesis of target molecules. They are described in this chapter beginning with hydrogen donors as radical traps.

B. Trapping with Hydrogen Donors

1. Mercury hydride

The reduction of alkylmercury salts **10** with hydrogen donors like $NaBH_4$ or Bu_3SnH leads to alkylmercury hydrides **11**. Although these hydrides have not yet been isolated, stereo-chemical,[2] polarographic,[3] and kinetic[4,5] studies indicate the existence of alkylmercury hydrides that trap alkyl radicals with rate coefficients of at least 10^7 (1/mol·s)[4] at room temperature. Therefore, high yields of C-C bond formation occur with very reactive alkenes like acrylonitrile, vinyl ketones, acrylates, fumarodinitrile, or maleic anhydrides.[6,7]

$$RHgX \xrightarrow{\ NaBH_4\ } R\,Hg\,H$$

10   **11**

Styrene, vinylidene chloride, and crotonic ester give low
yields[8] because they are too unreactive[1] to compete with
mercury hydrides. With these olefins, the amount of C-C bond
formation products can be increased by using tin hydrides as
hydrogen donors,[9] which react at least ten times slower with
alkyl radicals than mercury hydrides.[4,10] An even slower
hydrogen donor than a tin hydride is tributylgermanium
hydride[11] which can also be used in syntheses.[12]

$$C_6H_{11}X \quad + \quad \text{(alkene)}_Y \quad \longrightarrow \quad H_{11}C_6\text{-}Y$$

12   **2**   **13**

Alkene **2**   Yield of **13**

	Mercury Method[a]	Tin Method[b]
$\diagup\!\!\diagup$ CN	89	95
$\diagup\!\!\diagup$ CO$_2$CH$_3$	80	85
$\diagup\!\!\diagup$ C(=O)CH$_3$	60	85
CH$_3$O$_2$C$\diagdown\!\!\diagup$CO$_2$CH$_3$	55	78
$\diagup\!\!\diagup$ C$_6$H$_5$	40	83
$>\!\!=\!$CCl$_2$	35	90
H$_3$C$\diagdown\!\!\diagup$CN	45	85

a) C$_6$H$_{11}$HgOAc / NaBH$_4$;   b) C$_6$H$_{11}$I/ Bu$_3$SnH

The advantage of the mercury method over the tin method lies on its very mild reaction conditions. High temperature initiators or photolytic conditions are not needed because the mercury hydrides start the chain by spontaneous decomposition and the chain length is very long. The reactions, which require only a few minutes, can be carried out at room temperature in CH_2Cl_2 and there is no need to work under N_2.

Functional groups like halides, alkenes, cyclopropanes, ketones, boro- and metallorganic compounds can be used as precursors for radical C-C bond formation reactions because they react easily to give organomercury compounds.

a. Halides

Halides can be easily transformed into alkylmercury salts via Grignard compounds and in situ mercuration.[8,13]

$$RHal \xrightarrow{\text{Mg}} RMgHal \xrightarrow{\text{Hg Hal}_2} RHgHal$$

This route makes prim., sec., and tert. radicals accessible. A characteristic feature of radical reactions is that one can easily use a tert-butyl group in C-C bond formations.[14]

95   :   5

The yields are high because tert-butyl radicals are very nucleophilic (high lying SOMO) and, therefore, react very fast with electron poor alkenes.[1] The stereochemistry of the hydrogen abstraction step is influenced by the bulky tert-butyl group. For example, with methylmaleic anhydride the cis-isomer is formed predominately because in the π-radical **14** the tert-butyl group efficiently shields one side of the p-orbital.[14,15]

By similar reasoning, alkynes lead predominately to cis-alkenes in reactions with tert-butylmercury salts.[16]

b. Alkenes

Alkenes with an electron-rich terminal double bond **15** can be coupled with electron-poor alkenes **2** in a one pot reaction.[17] The reaction proceeds via hydroboration of **15** and transfor-

mation of the boron-carbon bond into a mercury-carbon bond,
which on reaction with NaBH$_4$ and alkene **2** gives product **16**.

These reactions tolerate substituents like Br, Cl, OH, OAc,
OTs, OR, and CO$_2$R. Even alkenes with two different double
bonds can be used. For example, vinylcyclohexene **17** reacts
with **18** exclusively to give product **19** because the hydroboration
step is faster at the terminal double bond, and prim. C-B
bonds are mercurated faster than sec. ones.[17]

The high regio- and stereoselectivity of this method is
demonstrated by the reaction of ß-pinene **20**, which gives
exclusively product **21**.[17]

20   **21**   53%

1) BH$_3$ / THF
2) Hg(OAc)$_2$
3) Na BH$_4$

The mercuration reactions of alkenes offer an even broader use
of this system. Mono-, di-,and trisubstituted alkenes **22** are
readily mercurated, according to the Markovnikov rule, with
mercury salts in the presence of the solvent HA. Thus,
alcohols are formed in the presence of water, ethers in
alcohols, esters in acids, and amides in acetonitrile.[13]
Combining the solvomercuration with radical C-C bond formation
reactions offers a broad variety of synthetic possibilities.[18]

22   **2**   **23**

1) Hg(OAc)$_2$/ HA
2) Na BH$_4$

A = OH,OR,OAc,NHAc

The synthesis can be carried out as a one-pot reaction:
After solvomercuration, the reaction mixture is diluted with
dichloromethane, the electron-poor alkene is added in about
three- to tenfold excess, and finally NaBH$_4$ or NaB(OCH$_3$)$_3$H
is added to start the radical process.[18] Barluenga[19] has shown
that it is often advantageous to carry out the synthesis as a

phase transfer reaction in water/dichloromethane using Triton B as a phase transfer catalyst. The introduction of the additional functional group A in this reaction sequence permits subsequent reactions, e.g. lactonization.[20]

Kozikowski[21] has used this method to synthesize the antibiotic malyngolide **25**, as a racemic mixture of diastereomers, from allyl alcohol **24**.

With cyclic alkenes, the stereochemistry of the radical
reaction can occur with high selectivity. Thus, Henning and
Urbach[22] have synthesized α-aminonitriles **28** starting from
cyclohexene, via solvomercuration, radical coupling with
α-chloroacrylonitrile (**26→27**), and intramolecular substitution,
which gave **28a** and **28b** in a 95:5 ratio.

$$95 \quad : \quad 5$$

The C-C bond formation reaction of the cyclic radical **29**, and
also the hydrogen abstraction of adduct radical **30** occur with
high diastereoselectivity. This is presumably due to the
preferred conformation of the radical **30**, which is attacked
from the less hindered side.

In the syntheses of **25** and **28,** two and three chiral centers are formed, respectively. Four chiral centers are introduced in the solvomercuration/reductive coupling sequence between galactal **31** and methylmaleic imide **33.**[6,23]

       31   **32**   **33**   **34 (55**

                    35   **36**

The four chiral centers in **34** are formed in three different steps:

- Electrophilic addition to an alkene.
- Radical addition to an alkene.
- Radical hydrogen abstraction from an alkylmercury hydride.

The mercuration of galactal **31** takes place *cis* to the complexing OAc substituents; subsequent *trans*-attack by the solvent CH_3OH leads to organomercuric salt **32.** Reduction of **32** gives carbohydrate radical **35** which is attacked at the equatorial side by methylmaleic imide **(35→36).**[23] In the cyclic radical **36,** the carbohydrate substituent shields the *cis*-side of the adjacent semioccupied p-orbital; therefore, the organomarcury hydride attacks from the *trans*-side giving rise to *cis*-product **34.** The diastereoselectivities of these reactions are higher than 95:5; only the diastereoface-differentiating reaction that generates the first chiral center at the maleic imide during the addition of **33** at **35** has a low selectivity of 55:45.

Alkenes **37**, with nucleophilic neighboring groups such as alcohols, carboxylates, amides, amines, and olefins, lead to cyclic products **38** in the mercuration/reductive coupling sequence.[24-26] The nucleophiles act as intramolecular traps for the mercurated olefins.

Educt **37**	Product **38**	Yield (%)
		62
		40
		41
		90

This reaction sequence was applied by Danishefsky for the synthesis of δ-coniceine **39** and indolizidinone **40**.[25]

In the synthesis of brevicomin **43**, the first step is the mercuration of the chloral adduct of allylalcohol **41**, which occurs with a 85:15 diastereoselectivity.[27] Reductive coupling with methyl vinyl ketone yields product **42**, the precursor of brevicomin **43**. It should be noted that iodation of the mercurated intermediate and C-C bond formation using Bu₃SnH gives higher yields of **42**.[28]

c. Cyclopropanes

Cyclopropanes **44** can be readily solvomercurated [13] to give acyclic organomercury salts **45**. The radical C-C bond forming reaction with, for example, methylvinyl ketone leads to 1,6-disubstituted products **46**. [29]

In analogy to the mercuration of alkenes, the mercury ion
attacks the lowest substituted C-atom of the cyclopropane, and
the nucleophile reacts from the *trans*-side with the highest
substituted cyclopropyl C-atom. Alkenes with electron-withdrawing substituents like CN, CO_2R, COR, Cl etc., and cyclopropanes with one to four alkyl or aryl groups can be used in
these syntheses.[30] In most cases, the reactions are carried
out without isolation of the organomercury salts.[30,31]

Different functional groups are introduced into the products
by varying the solvents.[32]

A :	OCH$_3$	OH	OAc	NHCOCH$_3$(CH$_3$CN)
yield :	82	62	65	65%

Also neighboring nucleophiles can trap the mercurated cyclopropane cation in an intramolecular reaction.[32]

The use of carbonyl compounds **47** as precursors for cyclopropanes broadens the application of this method considerably.
Thus, silylation of carbonyl compounds **47** gives **48**, which leads
to cyclopropanes **49** and C-C bond formation products **50**.[33]

The intermediate radicals **51** are, because of their nucleophilicity, analogues to homoenolate anions **52**.[34] But, whereas
anions **52** are mainly used in reactions with carbon-heteroatom
double bonds, radicals **51** attack carbon-carbon double bonds.
Both anions and radicals give products of umpolung of the
reactivity.

ROS-C

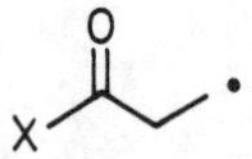

51 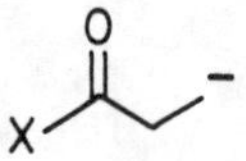   **52**

Aldehydes, ketones,and carboxylic esters can be used as carbonyl compounds. The examples shown below are one pot reactions
after the cyclopropanation step.[33]

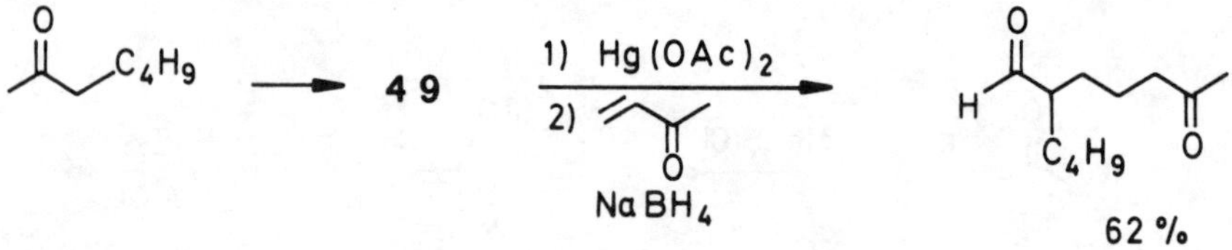

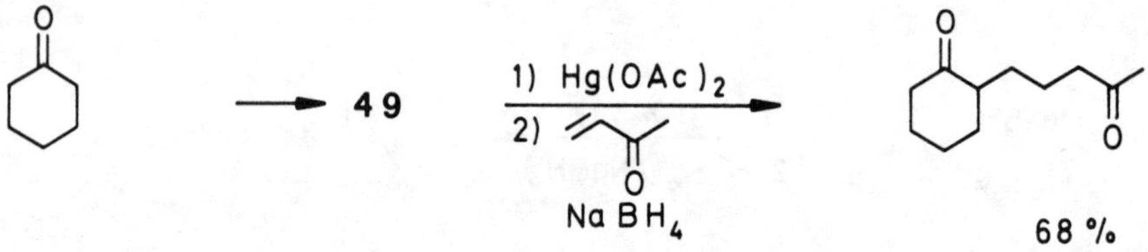

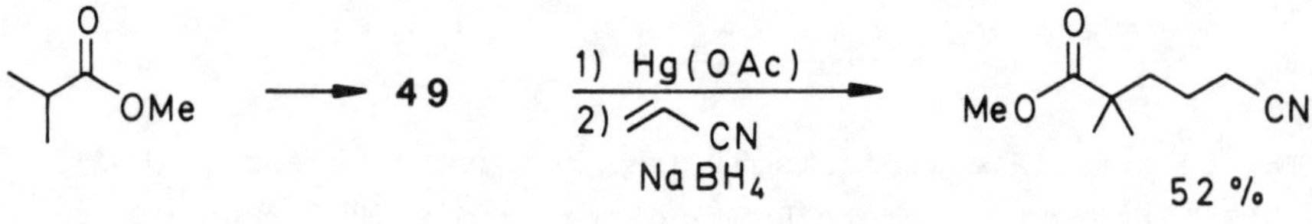

Solvomercuration of aldehydes in acetic acid leads, after
reaction with $NaBH_4$, to radicals **53**. However, solvomercuration
in alcohol or water/acetone gives radicals **54** which can
easily rearrange.[35]

53   **54a**   **54b**

Under special conditions (different stability of the radicals,
low concentration of the trapping alkene) one can also obtain
C-C bond formation products from the rearranged radical intermediates. The very fast rearrangement of radicals **54**, which
is on the order of $k = 10^6$ (s^{-1}) at 20°C,[36] makes this possible.

Starting from ketones, isomeric enolethers can be synthesized
by the method of House.[37] This leads to isomeric cyclopropanes and C-C bond formation products.[33]

The rearrangement of radicals **55**[38] and **56**,[39] generated from ketones and carboxylic esters, respectively, is too slow to compete with the C-C bond formation reaction.

55   **56**

Therefore, no rearrangement occurs when ketones and carbo-xylic esters are used as educts for the cyclopropane route.[33]

d.   Ketones

Ketones **57** can be transformed via hydrazones into α-substituted organomercury salts **58**,[13] which after reduction in the presence of alkenes, e.g. acrylonitrile, give products **59**.[40] A large variety of ketones can be used; even a tert-butyl group is tolerated.[41]

57   **58**   **59**

$R^1 = CH_3$; $R^2 = CH_3(70\%)$, $C_2H_5(65\%)$, $CH(CH_3)_2(62\%)$, $C(CH_3)_3(46\%)$

Again, alkenes with electron-withdrawing substituents give
good yields, as the reactions of norcampher **60** show.[41]

60

77 %   75 %   60 %   58 %   81 %   52 %

In these syntheses, C-C bonds between the electrophilic
C-atoms of the carbonyl groups and the alkenes are formed.
Therefore, this is another example of an umpolung reaction.
The umpolung of the reactivity occurs in going from the ketone
to the radical **61** which is a nucleophile, although the OAc
group reduces the nucleophilicity a little bit.[42]

6 1

2. Tin hydride

Since the work of Kuivila[43] it has been known that alkyl
radicals abstract hydrogen atoms from tin hydrides. Recently
it has been shown that trialkyl or triaryltin hydrides, for
example tributyltin hydride **62**, can also be used as traps for
adduct radicals **3**.[44,45] The resulting tin radicals, for
example **63**, don't decompose like the mercuric radicals, but
react with suitable precursors **64** to give back the starting
radical **1**.

Halides,[44,45] alcohols,[9] selenides,[44,46] and tert. nitro
compounds[47,48] can be used as radical precursors **64**.

R — Hal $\xrightarrow{\text{Bu}_3\text{Sn}\cdot}$ R •

R — OH $\longrightarrow$ $\underset{\displaystyle RO-\overset{\displaystyle S}{\overset{\|}{C}}-X}{}$ $\xrightarrow{\text{Bu}_3\text{Sn}\cdot}$ $\underset{\displaystyle R-O-\overset{\displaystyle S-SnBu_3}{\underset{\cdot}{\overset{|}{C}}}-X}{}$ $\longrightarrow$ R •

R — SePh $\xrightarrow{\text{Bu}_3\text{Sn}\cdot}$ R •

R — NO$_2$ $\xrightarrow{\text{Bu}_3\text{Sn}\cdot}$ $R-N\overset{O}{\underset{O\cdot}{\diagdown}}SnBu_3$ $\longrightarrow$ R •

For a successful synthesis, it is important that the tin
radicals react faster with the radical precursor **64** than with
the alkene **2**. Therefore, most alkyl chlorides, prim. and sec.
nitro compounds, and isonitriles, which are slowly reduced by
tin hydride, cannot be used in C-C bond forming reactions.

a. Halides

Reactions of halides with alkenes show that prim., sec., and
tert. bromides and iodides, and alkenes with electron-withdrawing or radical-stabilizing substituents can be successfully used.[9] The reactions are carried out either photolytically at room temperature or thermally with radical
initiators (e.g. AIBN). In some cases catalytic amounts of
tin hydrides or even tin halides with an excess of NaBH$_4$
can be used. Working with low concentrations of tin hydrides
lowers the amount of reduction products of starting material.

$$R-Hal \quad + \quad \diagup\!\!\diagdown^Y \quad \xrightarrow[\text{Bu}_3\text{SnCl / NaBH}_4]{\text{Bu}_3\text{SnH} \quad \text{or}} \quad R\diagdown\!\!\diagup\!\!\diagdown^Y$$

Alkylhalide	Alkene	Yield (%)	
		$h\nu$	Δ(AIBN)
$n-C_6H_{13}I$	⟋CN	80	
$c-C_6H_{11}I$	—"—	95	
$t-C_4H_9Br$	—"—	87	98
$n-C_6H_{13}Br$	—"—		68
$c-C_6H_{11}Br$	—"—		80
$t-C_4H_9Br$	—"—	62	
$c-C_6H_{11}I$	⟋C(=O)CH₃	85	
—"—	⟋CHO		90
—"—	⟋CO_2CH_3	85	
—"—	⟋C_6H_5	83	
—"—	⟋CCl₂ (Cl, Cl)	87	
—"—	NC⟍⟋CN		72
—"—	⟍⟋CN	86	

CO and CN bonds located in ß-position to the radical center remain unchanged during the C-C bond formation. Therefore, this approach offers a suitable way to transfer the chirality of the starting material into the products. Thus, molecules from the chiral pool, like lactic acid **66** or the protected glyceraldehyde **68** are suitable precursors for the chiral lactones **67** and **69**, respectively.[49]

For the synthesis of malyngolide **72** the chiral center is intro-duced by Sharpless epoxidation of allylic alcohol **70**.[50]

R = C$_9$H$_{19}$

Baldwin has shown that also aminoacids can be used as precursors for these radical reactions.[51]

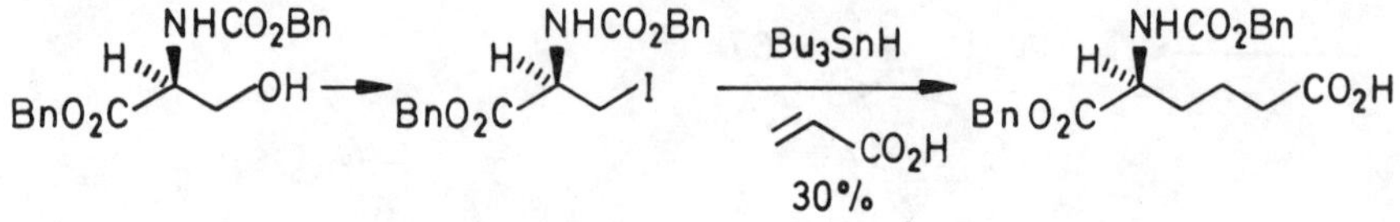

Because of the mild reaction conditions, radical C-C bond forming reactions have turned out to be suitable for the synthesis of C-glycosides.[45,52]

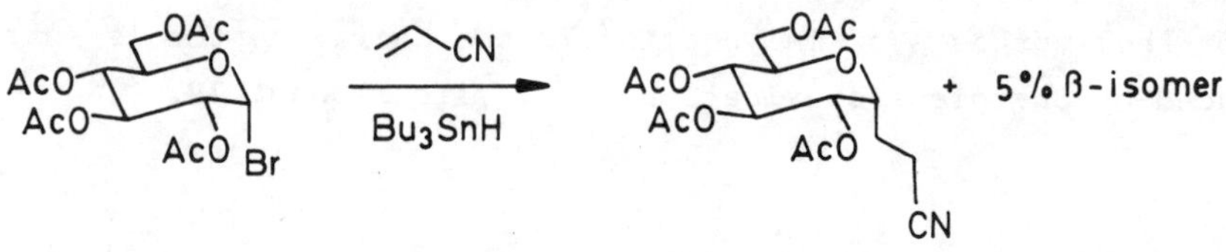

73 (70%)

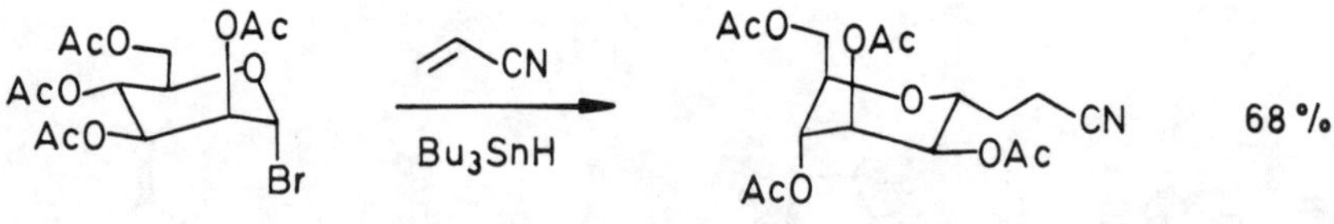

74

The predominant formation of α-C-glucosides **73** and **74** results
from equatorial attack at the boat conformation of the glucosyl
radical **75** and from the shielding effect of the axial
substituent in the chair conformation of the mannosyl radical
76, respectively.[46]

75   **76**

The glucosyl radical exists in the boat conformation because
of the interaction between the high SOMO energy of the alkoxyalkyl radical and the low LUMO energy of the C-O bond.[46,53]

Using this methodology C-disaccharide **78**, in which the oxygen atom between the pyranosyl rings of the disaccharide is substituted by a CH_2 group, is readily available from alkene **77**.[54]

77   **78**

CC-bonds can be easily synthesized not only at the anomeric center, but also at all of the other carbohydrate ring positions, for example at C-4.[9]

85   :   15

The newly formed C-C bonds in these C-branched sugars are predominately equatorial when the neighboring substituents are also equatorial.[55] Axial attack predominates only if both neighboring substituents are axial.[56]

R =			
R = H	65	:	35
CH$_3$	74	:	26
OAc	82	:	18
Li	96	:	4

The tin hydride method also allows the use of electrophilic radicals in C-C bond forming reactions. Thus, the reaction of chloromalonic ester with enolethers gives addition products, but with bromomalonic ester, substitution products are formed.[57]

$(H_5C_2O_2C)_2CHCl$ + $\diagup\!\!\diagdown OC_4H_9$ $\xrightarrow[60\%]{Bu_3SnH}$ $H_5C_2O_2C$ ... OC_4H_9 ; $H_5C_2O_2C$

$(H_5C_2O_2C)_2CHBr$ + $\diagup\!\!\diagdown OC_4H_9$ $\xrightarrow[75\%]{Bu_3SnH}$ $H_5C_2O_2C$ ... OC_4H_9 ; $H_5C_2O_2C$

This different behavior is the result of the different rates of halogen abstraction. In the competition system $XCH(CO_2C_2H_5)_2$/ Bu_3SnH, the hydrogen donor reacts faster for X=Cl, but the halogen donor is faster for X=Br. Subsequent HBr elimination then gives the substitution products.[57]

Chloroacrylonitrile can also be used as precursor for an electrophilic radical that reacts with enolether **79** to give deoxysugar **80**.[58]

79   **80**

b. Alcohols and selenides

Barton[59,60] has shown that xanthates and similar thioacylated alcohols can be deoxygenated with tin hydrides.

The attack of the tin radical at the sulfur atom is fast
enough for a successful application of this method for C-C bond
forming reactions. Thus, the C-branched deoxysugars **82** can be
easily synthesized from xanthate **81**.[9]

81   **82**

In the radical generated from **81**, both substituents shield
one side of the radical center; therefore the C-C bond forming
reaction occurs with high stereoselectivity. Selenides are
also suitable precursors.[46]

c. Nitro compounds

Nitro groups of sec. and tert. nitro compounds are very effectively replaced by hydrogen on treatment with tributyltin hydride.[61] Tight radical salts **83**[62] or nitroxide radicals **84**[48,63]
have been proposed as intermediates in these reactions.

$$R - NO_2 \xrightarrow{\text{Bu}_3\text{SnH}} R - H$$

$$R - N \overset{\displaystyle O}{\underset{\displaystyle O}{\cdot}} \quad {}^+SnBu_3 \qquad\qquad R - \overset{\displaystyle \overset{\cdot}{O}}{\underset{|}{N}} - O - SnBu_3$$

83   **84**

Tert. nitro compounds react with tributyltin hydride fast
enough for a successful application of this method for C-C
bond formation. Thus, **85** gives **86**, a C-glycoside of a keto
sugar.[47,52]

85   **86**

Ono[48] has shown that high yields in C-C bond forming syntheses
with nitro compounds need high temperatures, long reaction
times and large amounts of AIBN, because the radical chain is
not very long. Nevertheless, several functional groups are
tolerated under these conditions.

3. Germanium hydride

Trialkylgermanium hydride **87** is a less reactive hydrogen donor than trialkyltin hydride **62**.[11,64] But in reactions with alkyl halides **64** the germyl radical **88** is as reactive as the stannyl radical **63**.[11]

Therefore, in C-C bond forming syntheses with "unreactive" alkenes the amount of reduction product **89** should decrease by using germanium hydride instead of tin hydride.

Actually, Hershberger[12] has shown that cyclohexenone **90** gives the alkylated product **91** in good yields.

But this method needs long reaction times and one can often get similar results by working with low concentrations of tin hydride.[65]

Silanes are very unreactive hydrogen donors.[66] To build up a C-C bond forming chain reaction, high temperatures have to be used, and polymerization of electron-poor alkenes cannot be prevented.

4. "Carbon hydrides"

C-H bonds are much stronger than Hg-H, Sn-H, and Ge-H bonds.
Therefore, C-H bonds can be used as traps for adduct radical **3**
only if they are present in large excess and/or if substituents
reduce the C-H bond strength. In contrast to mercury, tin, and
germanium hydrides, an excess of the hydrogen donor **92** does
not lead to side-products because hydrogen abstraction by
educt radical **1** gives back the radical precursor **92**. A characteristic of this chain reaction is that hydrogen donation
to adduct radical **3** directly leads to **1**.

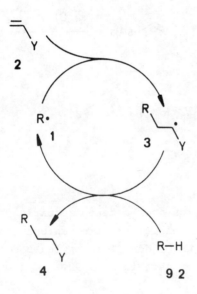

Because of their radical stabilizing effects, both electron
donating[67] and electron withdrawing groups[68] reduce the
strength of C-H bonds.[69] Thus, alcohols, ethers, acetals,
aldehydes, formic esters and amides, carboxylic esters, acids,
lactones, amines, amides, and lactams can be used as hydrogen
donors.[70] Under very special conditions even hydrocarbons add
to alkenes. In most of these reactions, the formation of

telomers plays a role and the chain lengths are relatively
short.

a. Alcohols, ethers, and acetals

Alkyl radicals attack alcohols at the α-C-H bond, because this
bond is much weaker than an O-H bond. The chain can be
initiated photolytically or thermally with initiators like
DTBP or Bz_2O_2. With electron-rich alkenes, the yields are not
very high although they often increase in going from methyl
via prim. to sec. alcohols.[71]

$$R^1\!\!\diagdown\!\!\underset{R^2\diagup}{}\!\!CHOH \quad + \quad \diagup\!\!\diagdown\!\!C_6H_{13} \quad \xrightarrow{\text{DTBP}} \quad C_8H_{17}CR^1R^2OH$$

$R^1 = R^2 = H \quad : 16\% \; ; \; R^1 = CH_3, R^2 = H : 28\%$

$R^1 = R^2 = CH_3 : 46\%.$

Addition to electron-poor alkenes requires photolytical
initiation with a sensitizer like benzophenone. Under these
conditions, Schenck[72] added isopropyl alcohol to maleic acid
in the synthesis of terebic acid **93**.

93

Fraser-Reid[73] introduced this method to carbohydrate chemistry and synthesized branched-chain sugars from enone **94**.

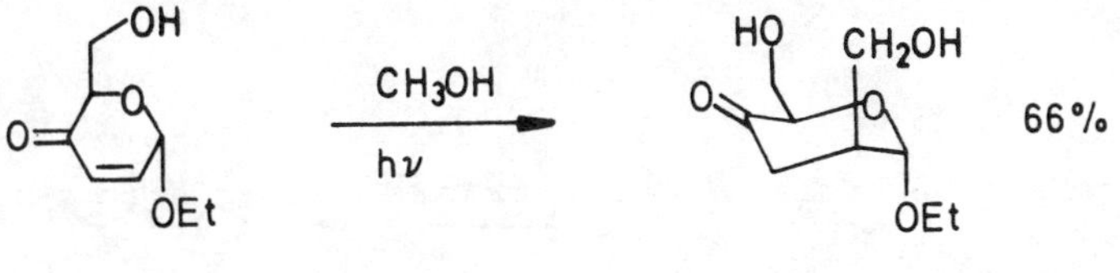

94

In addition reactions of ethers and acetals, electron-poor alkenes have been used predominately,[74,75] presumably because the educt radicals **95** have to be rapidly trapped to prevent ß-bond cleavage.[76]

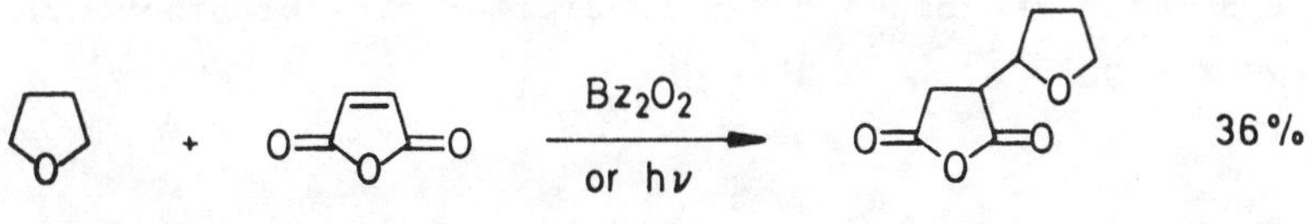

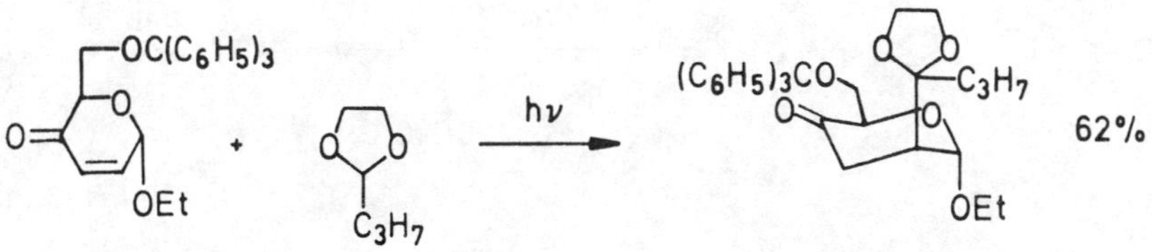

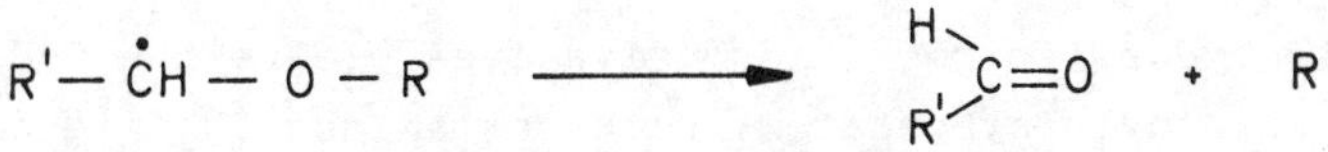

95

b. Aldehydes and formic acid derivatives

The radicals generated from aldehydes and formic acid derivatives are reactive σ-radicals which show nucleophilic
behavior.[77,78] The yields of addition products are often fairly
good, especially if electron-poor alkenes are used.[79]

41

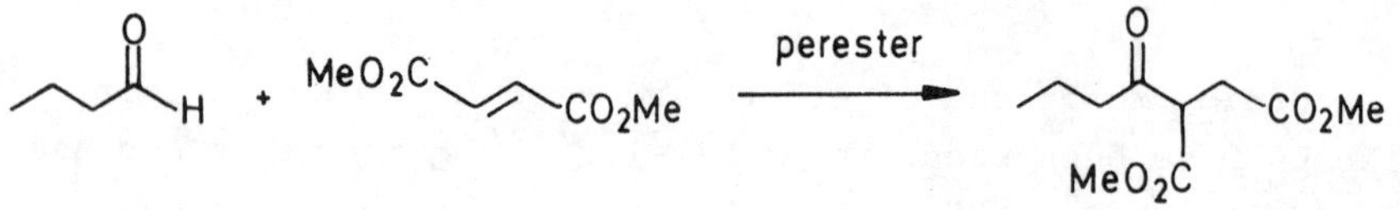

7(

The reactions can also be carried out under photolytical conditions with benzophenone as sensitizer[75] or with benzoyl-
perbenzoate 96 as initiator.[78]

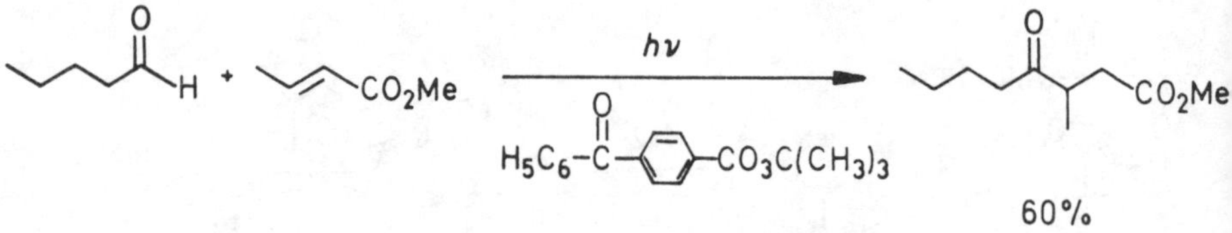

60%

96

Like aldehydes, formic acid derivatives cleave the C-H bond
of the carbonyl carbon atom. With formamides the yields are
often higher than with formic esters.[80]

$$H-C{\overset{\displaystyle O}{\underset{\displaystyle NMe_2}{}}} \quad + \quad \diagdown\diagup C_6H_{13} \quad \xrightarrow{\text{DTBP}} \quad C_8H_{17}CONMe_2 \qquad 60\%$$

$$H-C{\overset{\displaystyle O}{\underset{\displaystyle OMe}{}}} \quad + \quad \diagdown\diagup C_4H_9 \quad \xrightarrow{\text{DTBP}} \quad C_6H_{13}CO_2Me \qquad 20\%$$

c. Ketones

Via electrophilic radicals ketones undergo addition reactions with electron-rich alkenes. In order to obtain good yields of 1:1 adducts, the reactions must be initiated with di-tert-butyl peroxide.[81]

The presence of certain transition metal oxides can increase the yields.[82] Thus, without solvent, acetone and terminal alkenes in the presence of $Mn(OAc)_3$ afford mainly 1:1 adducts (see p. 89).[83]

d. Esters and lactones

With carboxylic esters and lactones, hydrogen abstraction can
occur either α to the carbonyl or at the alkoxy group, but
generally esters have been predominately alkylated at the
α-position of the acid moiety.[84]

$$C_6H_{13}CO_2Me \quad + \quad \diagup\!\!\!\diagdown C_6H_{13} \xrightarrow[55\%]{DTBP} C_8H_{17}\underset{C_5H_{11}}{CH}CO_2Me$$

For high regioselectivities and yields, esters with an
additional electron-withdrawing substituent should be used.
This leads to more electrophilic radicals which preferentially
react with electron-rich alkenes.[85]

$$MeO_2C\diagdown\!\!\!\diagup\!\!\!\diagdown CO_2Me \quad + \quad \diagup\!\!\!\diagdown C_6H_{13} \xrightarrow{DTBP} MeO_2C\diagdown\!\!\!\diagup\overset{C_8H_{17}}{\underset{CO_2Me}{}} \quad 67\%$$

$$\overset{CO_2Me}{\underset{CO_2Me}{|}} \quad + \quad \diagup\!\!\!\diagdown C_6H_{13} \xrightarrow{DTBP} \overset{MeO_2C}{\underset{MeO_2C}{}}\!\!\!\diagdown\!\!\!\diagup C_8H_{17} \quad 79\%$$

$$\overset{CO_2Et}{\underset{CN}{|}} \quad + \quad \xrightarrow{DTBP} \quad 56\%$$

Lactones give also addition products with high regioselectivity.[81,86]

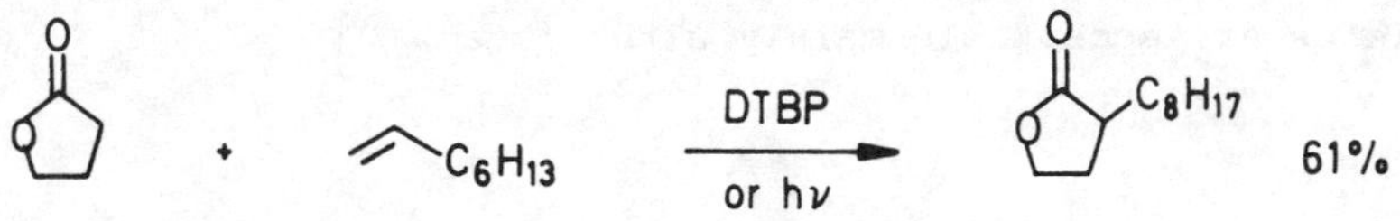

61%

e.   Amines

Amines react with electron-rich alkenes,[87] although the
intermediate aminoalkyl radicals are very nucleophilic. The
often observed high yields in these syntheses may be due to
the weak α-CH bonds of amines.

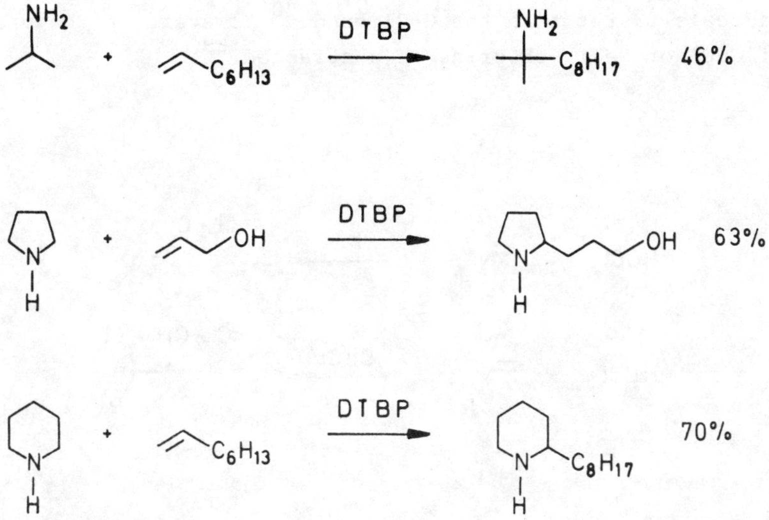

46%

63%

70%

f. Amides and lactams

In contrast to esters and lactones, amides and lactams react
predominately with the C-H bonds α to the nitrogen, presumably
because nitrogen stabilizes radicals very effectively. The
yields are acceptable mainly with lactams.[88]

major   minor

g. Chloroform

In contrast to other polyhaloalkenes, chloroform can act
as a hydrogen donor because the C-H bond is cleaved by alkyl
radicals faster than the C-Cl bond.[70] However, in the presence
of CuCl or $FeCl_2$ chlorine is abstracted.[89]

f. Hydrocarbons

Under high pressure and with high temperatures, even hydrocarbons can be used as hydrogen donors. The temperatures of
these reactions are so high (> 400°C) that no radical
initiators are needed. To avoid further reactions of the
products, the retention time in the hot zone should not be
too long. Best yields are obtained in reactions with electron-
poor alkenes.[90]

C_6H_{12} + (alkene with CN) $\xrightarrow{450°C}$ $H_{11}C_6$—(product with CN) 48%

C_6H_{12} + (alkene with OAc) $\xrightarrow{450°C}$ $H_{11}C_6$—(product with OAc) 10%

(isopropylbenzene) + (alkene with CN) $\xrightarrow{450°C}$ (product with CN) 30%

C. Trapping with Heteroatom Donors

1. Halogen donor

Light or peroxide initiated addition of polyhalogenated
compounds to alkenes was first observed by Kharasch.[91] In
these reactions, polyhalogenated alkyl radicals, for example
97, attack alkenes **98** to give adduct radicals **99**. Since the
C-X bond energies (X = Cl, Br, I) are low in polyhalogenated
compounds, radicals **99** are trapped by polyhaloalkanes **100**, to
give products **101** and the starting radicals **97**.[70,92]

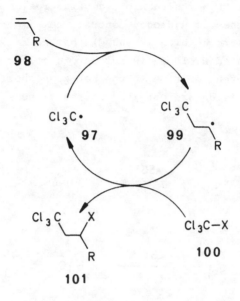

Halogenated radicals **97** are electrophiles[93] and react faster with electron-rich than with electron-poor alkenes. Bromides and iodides which are substituted by electron-withdrawing substituents like ester or nitrile groups can also be used in these syntheses.[94]

$$F_2C=CH_2 \quad + \quad C_2F_5I \quad \xrightarrow{h\nu} \quad I-C(F)(F)-CH_2-C_2F_5 \qquad 91\%$$

$$Me_2C=CMe_2 \quad + \quad BrCH(CN)_2 \quad \xrightarrow{h\nu} \quad (NC)_2CH-C-C-Br \qquad 98\%$$

Depending upon the rate of halogen abstraction, intramolecular reactions can compete with the intermolecular product forming step. For example, ß-pinene[95] gives the ring-opened product with CCl_4 and norbornadiene[96] leads to a mixture of bi- and tricyclic adducts. The ratio of the cyclized versus the non-cyclized products depends on the concentration of the halogen donor, the halogen atom X, and the reaction temperature.[97]

ß-pinene $+$ CCl_4 $\longrightarrow$ (cyclohexene with CCl_3 and Cl substituents) $\quad 97\%$

norbornadiene $+$ $XCCl_3$ $\xrightarrow[80\%]{AIBN}$ (bicyclic adduct with CCl_3, X) $+$ (tricyclic adduct with X, CCl_3)

This method was applied to the synthesis of permethrine acid 104. Addition of $XCCl_3$ (X = Br, Cl) to alkene 102 gives 103; base induced cyclization and elimination yields the target molecule 104.[98]

Reactions of perhaloalkenes can also be carried out with electron-poor alkenes in the presence of iron or copper salts. This approach has been used to synthesize permethrine acid 104 from acryloyl chloride.[99]

The addition product **105** is transformed into the ketene **106**, cycloaddition with isobutene, Favorskij rearrangement, and elimination then leads to **104**. The mechanistic details of the metal catalyzed addition reaction are not yet clear. It is, however, possible that an oxidative addition of Cu(I) to CCl_4 followed by a reaction with the electron-poor alkene occurs instead of a radical chain reaction.[100]

Kraus[101] has used tin salts of α-iodoacids and α-bromoacids for the synthesis of γ-lactones.

The reaction occurs presumably via addition of electrophilic radicals **107** to electron-rich alkenes and subsequent halogen abstraction (**108→109**). Cyclization then leads to the γ-lactone.

107   108   109

This mechanism seems to be more likely than cyclization of radical **108**.

2. Thio donor

A very useful synthetic method in which radicals are trapped by a thiocarbonyl group has been developed by Barton.[102]
Acid chlorides are converted with N-hydroxypiperidine-2-thione **110** to mixed anhydrides **111**, which are the radical precursors. Addition of electron-poor alkenes under either photolytical or thermal conditions leads to products **112** in which a C-C bond and a C-S bond have been formed.

In the reaction sequence   radical **1** is generated and attacks alkene **2** to give adduct radical **3**   that is trapped by the thio-compound **111**. Two successive ß-bond cleavages in **113** yield product **112** and the starting radical **1**.[103]

This reaction can be successfully carried out with a variety of different carboxylic acids as long as electron-poor alkenes are used.[102]

The reaction of nitroethylene with **111**, followed by oxidation, produces a carboxylic acid. Thus, in what could be a useful alternative to the classical Arndt-Eistert reaction, a carboxylic acid is transformed into the homologue.[104]

Radical C-C bond formation with 2-nitropropene and reductive cleavage with $TiCl_3$ leads to methyl ketones.[104]

Vinyl sulfones are also attractive alkenes because the adducts are versatile synthetic intermediates.[105]

Alcohols can be used as radical precursors via their half oxalates and the thio compound **114**. Thus, reactions with tert. alcohols lead to quaternary carbon centers.[106]

A side reaction, which involves the attack of alkyl radical 1 on the thio group of 111, leads to thioether 115.

111   **115**

Since 111 is a fast radical trap, this C-C bond forming synthesis gives high yields only with very reactive alkenes.

3. Borane donor

Brown[107] has shown that trialkylboranes are suitable precursors for alkyl radicals in C-C bond forming chain reactions with α,ß-unsaturated ketones and aldehydes. Adduct radicals 117 are generated by the addition of radicals 1 to alkene 116. They are then trapped by trialkylboranes to give, in a synchronous displacement or via intermediate 118, the starting radical 1 and 119, which is hydrolysed to give product 120.

The trialkylboranes are generated by in situ hydroboration of alkenes. The reaction of trialkylboranes with O_2 produces peroxiboranes, and the radicals generated from these peroxiboranes start the chain.[107,108]

Mannich bases derived from cycloalkanones, quaternized in situ, react smoothly with trialkylboranes in alkaline solution, presumably via a methylene ketone, to give the products of a C-C bond forming reaction.[109]

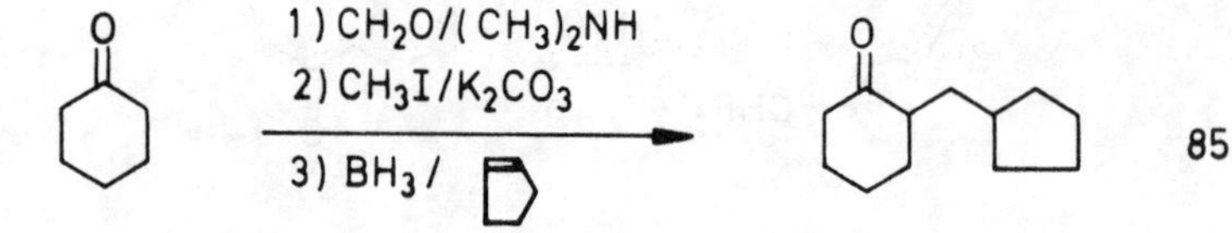

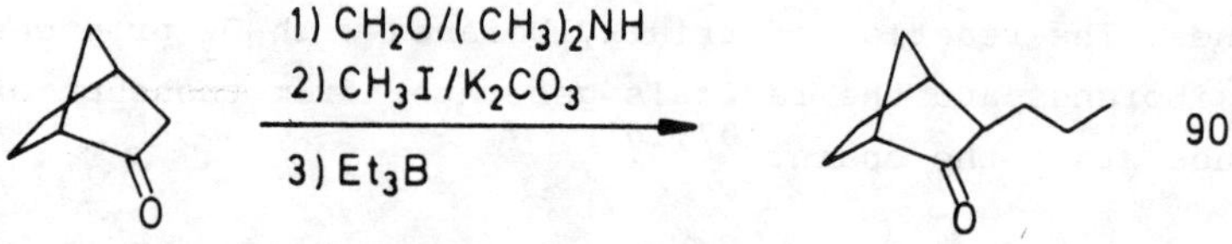

$$\begin{array}{c} 1)\,CH_2O/(CH_3)_2NH \\ 2)\,CH_3I/K_2CO_3 \\ 3)\,BH_3/ \end{array}$$

85 %

$$\begin{array}{c} 1)\,CH_2O/(CH_3)_2NH \\ 2)\,CH_3I/K_2CO_3 \\ 3)\,Et_3B \end{array}$$

90 %

Also α,ß-unsaturated ketones and aldehydes with an alkyl group
at the attacked C-atom of the double bond can be used in these
syntheses. However, the reactions are successful only if
enough radical initiators are present to restart the chains,
which have short length because of the less reactive
alkenes.[110]

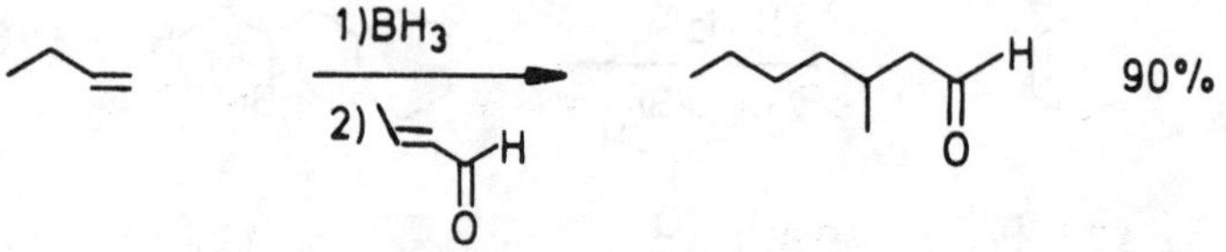

1)BH_3

2)

90 %

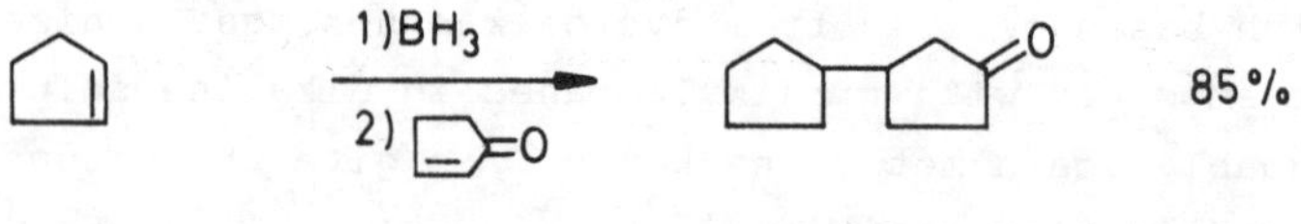

1)BH_3

2)

85 %

These syntheses are limited to α,ß-unsaturated aldehydes, ketones,and epoxides;[107] with ester or nitrile groups, the intermediate adduct radical cannot be trapped fast enough by the trialkylborane and polymerization occurs.

D.   Trapping by Electron Transfer Reactions

1.   Oxidation by metal ions

Cu^{2+}, Mn^{3+}, and Ce^{4+} ions are very efficient redox traps for nucleophilic alkyl radicals 122 [89,111] generated by addition of radicals 121 to alkenes 98. These metal salts oxidize radicals with high lying SOMOs to cations 123, which further react to give products, for example 124.

121   98   122

123   124

Since only nucleophilic radicals are oxidized fast enough, these syntheses are limited to electrophilic educt radicals **121** and electron-rich alkenes **98**. Acid, ester, acyl, nitro, and nitrile groups have been used as electron-withdrawing substituents X at the radical center of **121**. The preferred method for the generation of electrophilic radicals **121** involves the oxidation of hydrocarbons **125**, presumably via their tautomers **126** and the radical cation **127**.

$$X-\overset{\displaystyle |}{\underset{\displaystyle |}{C}}-H \quad \rightleftharpoons \quad HX=C{<} \quad \xrightarrow{Mn^+} \quad HX-\overset{+}{\overset{\bullet}{C}}{<} \quad \xrightarrow{-H^+} \quad X-\overset{\bullet}{C}{<}$$

125   **126**   **127**   **121**

$$X = CO_2H, \ CO_2R, \ COR, \ CN, \ NO_2$$

Heiba[112] developed this method for the synthesis of γ-lactones, which are formed by intramolecular trapping of the cations **123**.

Chlorinated radical precursors give α,β-unsaturated γ-lactones after HCl-elimination, and methylene lactones can be synthesized from cyanomalonic acids.[113]

ClCH₂CO₂H + $\overset{}{\diagup}$C₈H₁₇ $\xrightarrow[\text{52\%}]{\text{Mn}^{3+}}$ (structure) $\xrightarrow{\text{72\%}}$ (structure)

NCCH₂CO₂H + $\overset{}{\diagup}$C₈H₁₇ $\xrightarrow[\text{69\%}]{\text{Mn}^{3+}}$ (structure) $\xrightarrow[\text{2)MeI/NaHCO}_3]{\text{1)H}_2/\text{Pd}}$ (structure)
$$73\%$$

Temperatures of these reactions are often very high, but they
can be reduced to 25°C if electron-withdrawing groups α to the
carboxylic acid functions are present. This is, for example,
the case with ß-keto,[114] cyanoacetic, and malonic acids.[115]

(cyclohexene) + NCCH₂CO₂H $\xrightarrow[\substack{\text{25°C}\\\text{78\%}}]{\text{Mn}^{3+}}$ (structure)

H₅C₆$\diagup\diagup$OAc + NCCH₂CO₂H $\xrightarrow[\substack{\text{25°C}\\\text{77\%}}]{\text{Mn}^{3+}}$ (structure)

Ketones and phenyl rings can also trap the intermediate cations.[114]

If the nucleophilic groups don't react fast enough, the cations are stabilized by deprotonation.[116]

These C-C bond forming syntheses are not chain reactions because the Mn^{3+} ions are used up without being reoxidized. However, the reduced metal ions can be oxidized if peroxides are used as radical precursors. Thus, Minisci[89] generated radicals **1** from peroxides. Addition of **1** to electron-poor alkenes **2** gives adduct radicals **3** that are not oxidized because of the electron-withdrawing substituent Y and are therefore trapped by electron-rich alkenes **98**. Oxidation of **128** then gives cation **129** that reacts with electrophiles.

peroxide + Cu^+ $\longrightarrow$ $R^{\bullet}$ + Cu^{2+}

1

$R^{\bullet}$ + \ce{alkene}Y $\longrightarrow$ R\~\~$^{\bullet}$Y $\longrightarrow$ R\~\~Y$^{\bullet}$R'

1   **2**   **3**   **128**

98

$\xrightarrow[-Cu^+]{Cu^{2+}}$ R\~\~Y$^+$R'

129

A synthetic application of this reaction sequence is the formation of γ-lactone **130** from cyclohexanone, acrylic acid, and α-methylstyrene.[117]

$+ H_2O_2 + $ \ce{CH2=CHCO2H} $ + $ \ce{CH2=C(CH3)C6H5} $\xrightarrow[\substack{CH_3OH \\ 55\%}]{Fe^{2+}}$ HO_2C\~\~\~C_6H_5

130

The reaction proceeds via the peroxide 131 which, after
cleavage of the O-O bond, gives the nucleophilic alkyl radical
132.

131   132

2. Anodic oxidation

Anodic oxidation of anions 133 with electron-withdrawing
substituents such as ester, acyl, nitrile, or nitro groups
yields electrophilic radicals 134 that are resistent to
further oxidation and react with enolethers. The so formed
nucleophilic radicals 135 undergo rapid oxidation to cations
136 that are trapped either intra- or intermolecularly.[118]

133   134   135   136

The products are formed in yields of about 40%, and even
styrene can be used as alkene to trap the electrophilic
radicals.[118,119]

$(MeO_2C)_2\bar{C}H$ + $\diagup\!\!\!\diagdown OEt$ $\xrightarrow[\substack{MeOH \\ 37\%}]{-e-}$ $MeO_2C-\overset{\displaystyle MeO_2C}{\underset{}{C}}\diagup\diagdown\overset{\displaystyle OEt}{\underset{OMe}{C}}$

$Me_2\bar{C}NO_2$ + $\diagup\!\!\!\diagdown C_6H_5$ $\xrightarrow[\substack{MeOH \\ 43\%}]{-e-}$ $O_2N\diagup\diagdown\overset{}{\underset{OMe}{C}}\diagdown C_6H_5$

3. Syntheses via $S_{RN}1$ reactions

Alkyl radicals rapidly[120] attack nitroalkane anions, for
example 137, in a reaction that was first studied by Kornblum[121] with substituted benzyl radicals 138. The reaction
passes through the radical anion 139 which transfers an
electron to the starting halide 140 to yield product 141 and
the radical anion 142.[121,122] Because 142 loses a halide ion
in a monomolecular reaction step, the mechanism was termed
"unimolecular nucleophilic radical substitution" ($S_{RN}1$).[123]

$(CH_3)_2C=NO_2^-$

137

$ArCH_2C(CH_3)_2NO_2^{-\bullet}$

139

$ArCH_2^{\bullet}$

138

$ArCH_2X$

140

X^-

$ArCH_2X^{-\bullet}$

142

$ArCH_2C(CH_3)_2NO_2$

141

Alkylnitro radical ions can also be generated by reaction of nitro compounds with carbanions,[121] and Vasella[124] demonstrated the use of this method in carbohydrate chemistry.

Russell[125] has observed that alkylmercury salts react with nitrile, ester, acyl, and aryl substituted anions in aliphatic C-C bond formation reactions. The photoinitiated chain reaction occurs presumably via radical anions **143** and **144**.

The yields of several anions in their reactions with, for example, tert-butylmercury chloride are sufficiently high for a synthetic application of this methodology.[125,126]

$$(CH_3)_3CHgCl \quad + \quad R^- \quad \xrightarrow{h\nu} \quad (CH_3)_3C-R$$

$Me_2\bar{C}NO_2$ (69%) $H_2\bar{C}NO_2$ (68%) $(C_6H_5)_2\bar{C}CN$ (48%)

Fluorenyl$^-$ (44%) $C_6H_5\bar{C}(CO_2Et)_2$ (43%) $\bar{C}H_2COC_6H_5$ (54%)

E.   Fragmentation

After construction of the C-C bond by radical addition to alkenes **145** the adduct radicals **146** can be transformed into non-radical products not only by intermolecular trapping reactions, but also by splitting off a radical, for example, in a ß-bond cleavage.

$$R\bullet \quad + \quad \diagup\!\!\diagdown\!\!\diagup X \quad \longrightarrow \quad R\diagup\!\!\diagdown\!\!\diagup\!\!\diagdown X \quad \longrightarrow \quad R\diagup\!\!\diagdown\!\!\diagup \quad + \quad X\bullet$$

1   **145**   **146**

In these synthetic methods, an intramolecular reaction prevents adduct radical **146** from reacting with alkenes **145**. If this fragmentation is fast enough, then radicals **1** and **146**

need not to be of different polarity; that is, even electron-
rich alkenes **145** can be used for C-C bond forming reactions
with nucleophilic radicals. Various synthetic methods have
been developed during the last years in which different radicals are split off in distinct reaction steps.

1. Cleavage of carbon-heteroatom bonds

a. Carbon-tin bonds

Allylstannanes **147** have been used by Keck[127] for the synthesis
of C-C bonds via addition reactions with alkyl radicals **1**.
These syntheses benefit from the rapid cleavage of a C-Sn bond
ß to a radical center. Therefore, adduct radicals **148** give
allylsubstituted products **149** by splitting off a tributyltin
radical **63** that reacts with **64** to give **1**. Compound **64** can be
a halide, xanthate, thioether, or selenide.

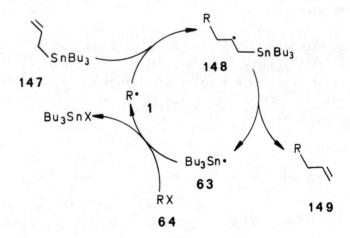

The radical chain is initiated by thermolysis of AIBN or
irradiation. A variety of precursors **64**, including carbohydrate
derivatives, can be used.[128,129]

The advantage of this allyltin method is that hydrogen or heteroatom donors, which often also react with educt radical **1** to give by-products, are not needed. Furthermore, the side reaction involving the addition of tin radicals **63** to the alkenes destroys neither **147** nor **63**, because they are immediately regenerated by ß-bond cleavage of **148**.

Therefore, in syntheses with allyltin **147**, less reactive radical
precursors like alkylchlorides and thioethers can be used.[128]

$$X = Cl, SC_6H_5$$

One application is the synthesis of pseudomonic acid C **152**
from L-lyxose **149** in which the C-C bond forming step is
carried out via radical allylation of **150**.[130]

149   **150**   **151**

152

Compared to electron-poor alkenes like acrylonitrile, olefin
147 reacts slower with nucleophilic radicals. Therefore, intramolecular reactions, for example the acetyl migration 153→154,
can compete with the intermolecular trapping reaction.[131]

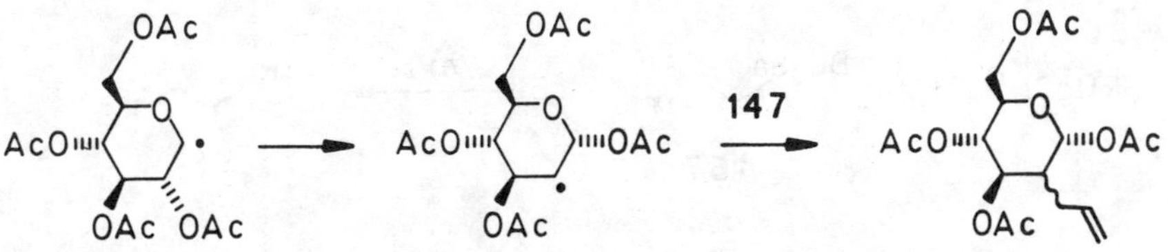

153   154

Unfortunately, this allylation method seems to be limited to
allyl- and methallylstannanes. With crotylstannane 155, for
example, the synthesis fails because the alkene is too
unreactive, and allylic hydrogen abstraction occurs.[128]

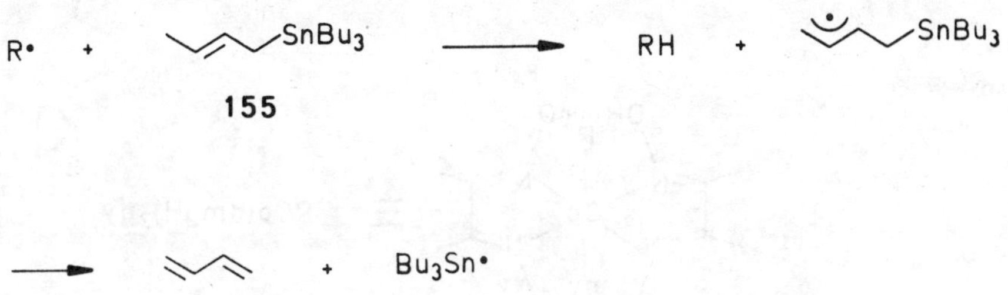

155

Baldwin[132] and Russell[133] have used vinylstannanes 156 and 157
for C-C bond forming reactions. The problem with these methods
is that the alkyl radicals have to attack the tin-substituted
olefinic C-atom in order to give products via ß-elimination
of radical 158.

$$R-Br \quad + \quad \underset{\underset{156}{Bu_3Sn \quad CO_2Et}}{\overset{=}{\diagup\!\!\diagdown}} \quad \xrightarrow[h\nu]{(Bu_3Sn)_2} \quad R\diagup\!\!\diagdown CO_2Et$$

$$RHgCl \quad + \quad \underset{157}{Bu_3Sn\diagup\!\!\diagdown\!\!\diagdown X} \quad \xrightarrow{h\nu} \quad R\diagup\!\!\diagdown\!\!\diagdown X$$

$$R^\bullet \quad + \quad Bu_3Sn\diagup\!\!\diagdown\!\!\diagdown X \quad \longrightarrow \quad \underset{\underset{158}{Bu_3Sn}}{R\diagup\!\!\diagup\overset{\bullet}{\diagdown}X} \quad \longrightarrow \quad R\diagup\!\!\diagdown\!\!\diagdown X$$

b. Carbon-cobalt bonds

Dissociation energies of C-Co bonds in organocobaloximes **159** are on the order of 20 kcal/mol for benzyl substituted compounds.[134]

$$\equiv \quad RCo(dmgH)_2py$$

159

A radical center in ß-position leads to a rapid cleavage of the C-Co bond. Therefore, radical **161**, generated by addition

of **1** to allylcobaloxime **160**, yields product **149** and the
cobaloxime radical **162**. This radical abstracts a halogen from
a suitable precursor and gives back the starting radical **1**.[135]

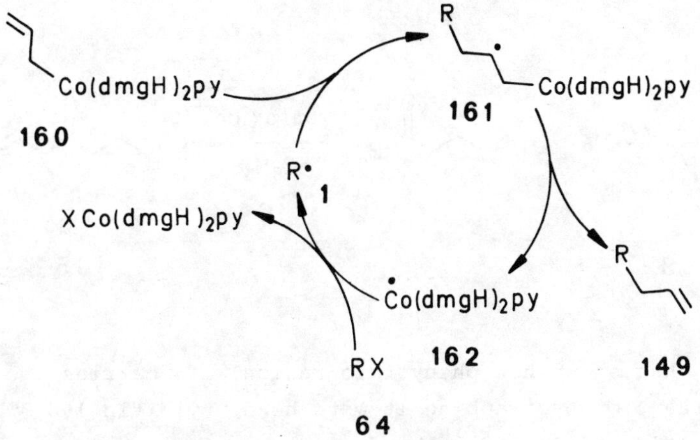

64

Syntheses with allylcobaloximes have been carried out mainly
with electrophilic radicals. In contrast to the allyltin
method, crotyl and higher alkylated systems can also be
used.[136]

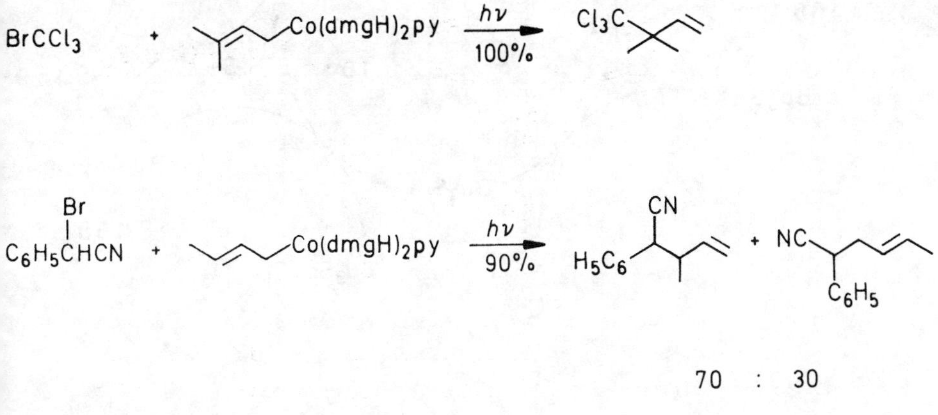

70 : 30

c. Carbon-sulfur bonds

Alkyl radicals with ß-C-S bonds undergo rapid ß-scission
reactions. Thus, butyraldehyde adds to **163** and yields, as a
side-product **165** after ß-elimination of phenylthio radical
164.[137]

163   **165**

Keck[138] has shown how phenylthio radical **164** carries the
chain reaction by trapping it with hexabutylditin **168**. The
tributyltin radical **63** thus formed reacts with **64** and
generates the starting radical **1**.

Halides and selenides are suitable precursors for educt radicals **1**.[138]

Barton[139] used thio compound **111** to trap tert-butylthio radical **171**. Radical **1**, thus generated, adds to **169** and gives product **149** via **170**.

These syntheses can be carried out as one-pot reactions in boiling chlorobenzene.[106,139]

2. Cleavage of oxygen-oxygen bonds

Peresters can undergo radical induced decomposition.[140] Maillard[141] used this for the synthesis of γ-lactones **174** from homoallylperesters **172** via radical **173**. The tert-butoxy radical **175** regenerates the alkyl radical **1** by hydrogen abstraction from RH.

Hydrocarbons, ethers, ketones, nitriles, and even methylene chloride have been used as radical precursors.[141]

$$R-H \quad + \quad \diagup\diagdown\diagup\diagdown CO_3Bu^t \quad \xrightarrow{Bz_2O_2} \quad R\diagup\diagdown\overset{O}{\diagup}\diagdown{=}O$$

$R = C_6H_{11}(73\%)$, $THP(65\%)$, $C_2H_5\underset{\underset{O}{\|}}{C}CHCH_3(58\%)$

$CHCl_2(70\%)$, $CH_2CN(64\%)$

In an analogous reaction,[142] epoxides are formed from allyl-peroxides.[141]

$$RH \quad + \quad =\diagup\diagdown O_2Bu^t \quad \xrightarrow[20-75\%]{Bz_2O_2} \quad R\diagdown\diagup\triangle$$

F. Miscellaneous Radical Addition Reactions

In the preceding sections of this chapter, mainly alkenes have been used as educts in the C-C bond formation reactions. For synthetic purposes alkenes are the most important traps, but 1,3-dienes or alkynes can also be utilized. The rates of radical addition to 1,3-dienes are slightly higher than those to alkenes.[143] However, since allyl radicals are the intermediates, mixtures are formed in the trapping reactions, although 1,4-adducts **176** are the main products in most cases.[70]

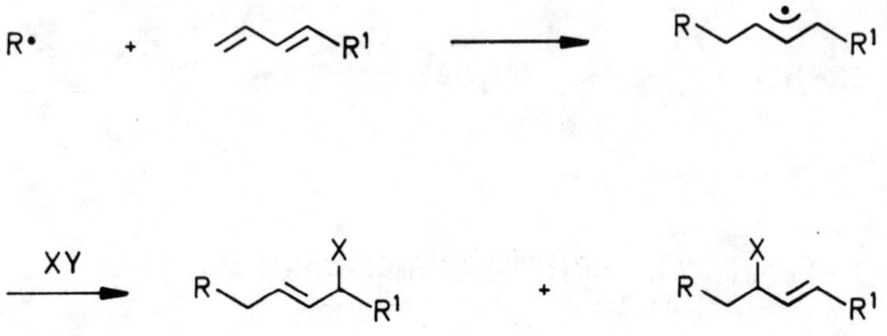

176

Alkynes are less reactive than alkenes.[1] Trapping of the intermediate vinyl radicals results in mixtures of *cis*- and *trans*-products. Since the semi-occupied orbital *trans* to substituent R is less shielded, *trans*-attack occurs faster, although the less stable *cis*-compound **177** is formed.[16]

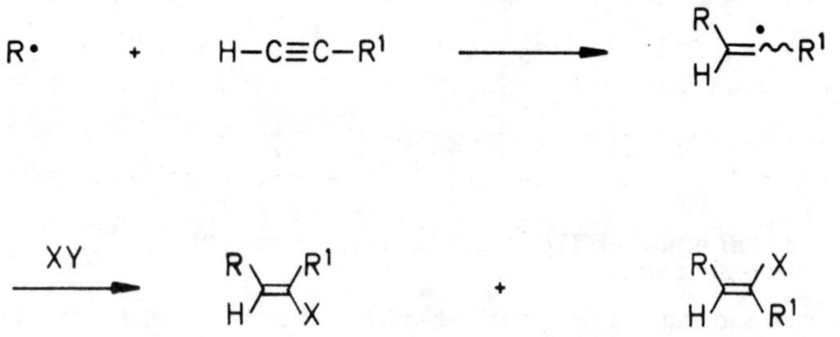

177

Allenes have been used in a few cases for intermolecular C-C bond forming reactions,[144] even the strained σ-bonds of cyclopropane **178** can trap alkyl radicals.[145]

NC–C(–Br)–CN + ==< —hν→ Br–CH₂–C(=C(CH₃)₂)–CH(CN)(CN)   77 %

Cl₃CCN + ==\⟨O–Co(dmgH)₂py⟩ —hν→ Cl₂C(CN)–CH₂–C≡CH   90 %

CH₃COCH₃ + **178** —hν→ (bicyclic product with –CH₂–CO–CH₃)   42 %

Carbon-oxygen π-bonds are as strong as carbon-carbon σ-bonds; thus, carbonyl groups react only slowly with alkyl radicals.[146] Furthermore, alkoxy radicals undergo rapid ß-bond cleavage. Nevertheless, several attempts have been made to use aldehydes and ketones as radical traps. Success can only be expected if the addition is favored by steric and polar effects. Thus, formaldehyde[147] and biacetyl[148] give products with moderate yields, but large amounts of peroxides were needed to restart the chain reactions.

$$R^{\bullet} \;+\; {>}C{=}O \;\rightleftharpoons\; R{-}\overset{|}{\underset{|}{C}}{-}O^{\bullet}$$

$$CH_3OH \quad + \quad CH_2O \quad \xrightarrow[30\%]{DTBP} \quad HO-CH_2CH_2-OH$$

$$R-H \quad + \quad \overset{\displaystyle O}{\underset{\displaystyle O}{\big\|}} \quad \xrightarrow[30-60\%]{DTBP} \quad R\overset{O}{\overset{\|}{-}}$$

Aldoximes,[148] substituted by an acyl group, gives higher yields and diethyl mesoxalate[149] leads to products even under irradiation.

$$\text{tetrahydrofuran} + \text{CH}_3\text{C(O)CH=NOH} \xrightarrow{(ROCO_2)_2} \text{product} \quad 67\%$$

$$\text{cyclohexane} + \text{CH}_3\text{C(O)CH=NOH} \longrightarrow \text{product} \quad 61\%$$

$$C_6H_{12} \quad + \quad EtO_2C-\overset{\displaystyle O}{\overset{\|}{C}}-CO_2Et \quad \xrightarrow[50\%]{h\nu} \quad C_6H_{11}\overset{OH}{\underset{|}{C}}(CO_2Et)_2$$

Stork[150] used tert-butylisonitrile in a multistep synthesis even though isonitriles react relatively slowly[151] with alkyl radicals.

Many organometallics **179** (e.g. M = Li[152], Mg[153], Zn[154], Cu[155], Co[156]) react with alkenes **2** to form addition products **4**.

In most of these reactions alkyl radicals are not involved. But Scheffold[157] has shown that the vitamin B_{12} catalyzed reaction of **64** with alkene **2** occurs via radicals which might be still in tight contact with the cobalt atom of vitamin B_{12}.

R—X + ⟍⟋Y $\xrightarrow[\text{solvent}]{\substack{e- \\ B_{12}}}$ R⟍⟋⟍Y

64   **2**   **4**

The reaction occurs via cleavage of the Co-C bond of alkyl cobalamine **180** leading to radical **1**, which is trapped by alkene **2**. Abstraction of a hydrogen atom, perhaps from the solvent, yields product **4**. The alkyl cobalamine **180** is regenerated by reduction (**181→182**) and alkylation (**182→180**).

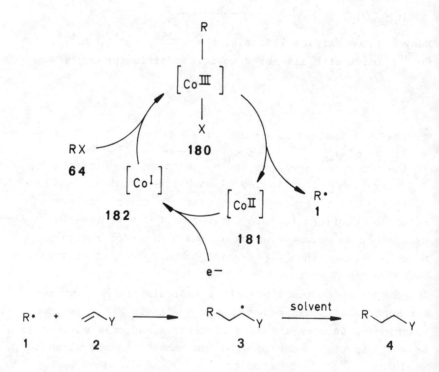

As alkyl radical precursors alkyl halides[157,158] can be used, acid anhydrides[159] lead to acyl radicals. The reduction is carried out either at the cathode or with Zn, irradiation facilitates the Co-C bond cleavage.

G. Radical-Radical Reactions

In the preceding sections of this chapter the formation of C-C bonds took place in reactions between radicals and non-radicals. It is, however, also possible to use radical combination reactions to form C-C bonds. For a successful synthesis, it is important to work in the absence of reactive molecules other than those which form the combining radicals.

1. Oxidative coupling

a. Oxidation of anions

Kolbe electrolysis of carboxylates leads to radicals which yield dimers by recombination.[118,160] Substituents such as esters,[161] ketones,[162] and fluorides[163] are tolerated in these syntheses; polyhalogenated[164] and allylic acids[118,165] can also be used.

$$R-CO_2^- \xrightarrow[-CO_2]{-e^-} R^\bullet \longrightarrow R-R$$

$$X(CH_2)_nCO_2^- \xrightarrow{-e^-} X(CH_2)_{2n}X$$

$X = CH_3,\ CO_2R,\ COR,\ OAc,\ F;\quad n = 4 - 16 : 45 - 95\%$

$$FCCl_2CO_2^- \xrightarrow[45\%]{-e^-} FCCl_2-CCl_2F$$

With allylcarboxylic acids, the configuration at the alkene is retained to a large extent, but the radicals recombine from both sides of the allylic system.[118]

Kolbe electrolysis has been used in a variety of syntheses, for example, in the synthesis of a pentacyclosquallene[166] or α-onocerin.[167] Dimerization of **183** and **184**, respectively, are the main steps.

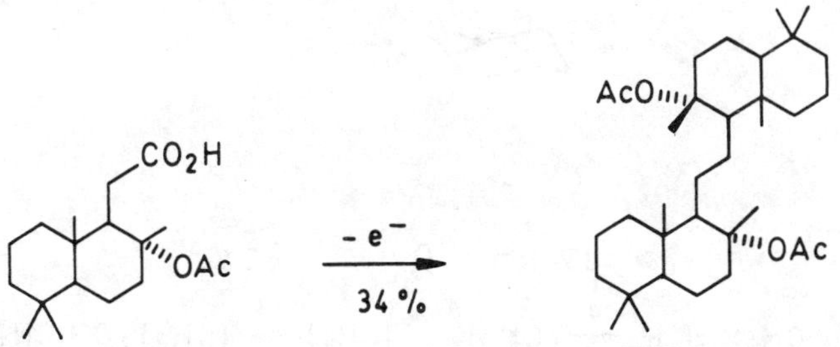

183

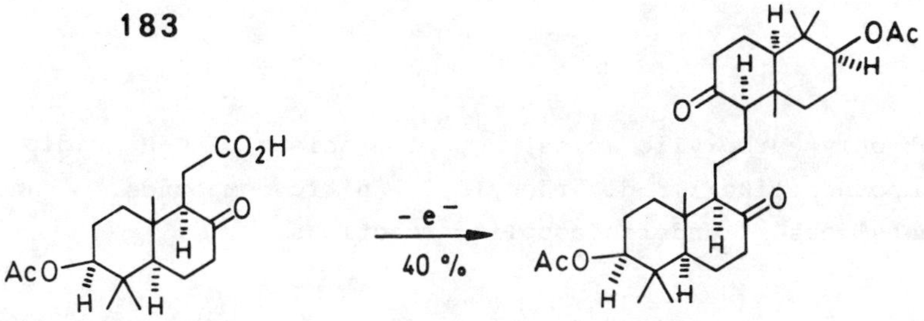

$$\xrightarrow[34\%]{-e^-}$$

184

$$\xrightarrow[40\%]{-e^-}$$

Mixed Kolbe electrolysis leads to unsymmetrical compounds, whose yields are increased if an excess of one of the acids is used. Schäfer[168] has synthesized pheromones under these conditions.

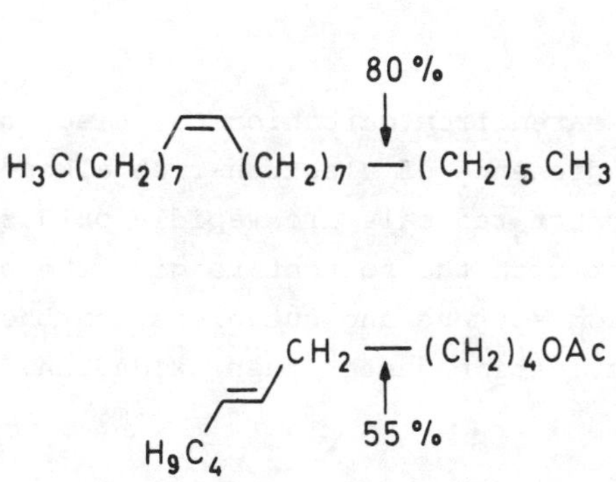

$$H_3C(CH_2)_7 \overset{}{\diagup}(CH_2)_7 \xrightarrow[]{80\%} (CH_2)_5 CH_3$$

$$H_9C_4 \overset{}{\diagup} CH_2 \xrightarrow[55\%]{} (CH_2)_4 OAc$$

$$33\% \rightarrow$$

$$48\% \qquad \qquad 62\%$$
$$CH_3(CH_2)_6 \longrightarrow (CH_2)_2 \qquad (CH_2)_2 \longrightarrow (CH_2)_2 CH(CH_3)_2$$

Not only carboxylic acids, but also anions of C-H acidic compounds like 1,3-dicarbonyls,[169] nitro compounds,[170] and acetylenes[171] undergo coupling reactions.

$$O_2N\overset{-}{C}H_2 \xrightarrow{-e^-} O_2NCH_2CH_2NO_2$$

$$RC\equiv C^- \xrightarrow{-e^-} RC\equiv C-C\equiv CR$$

The radicals generated from carbanions by electrolysis can be trapped in the presence of electron-rich alkenes. With enolethers, the adduct radicals are rapidly oxidized to cations which react with the solvent to give the products.[119] Adduct radicals from styrene and butadiene, on the other hand, undergo dimerization faster than oxidation.[172]

$$MeO_2C\diagup CO_2^- \ + \ \diagup\!\!\!\diagdown C_6H_5 \ \xrightarrow[38\%]{-e^-} \ $$

Electron-rich alkenes, like enolethers, yield oxidized dimers
presumably via recombination of radical cations **185** and
subsequent reaction with alcohol.[173]

Phenylsubstituted or 1,1-dialkylated alkenes can also be used for these reactions.[118] Coelectrolysis of different alkenes gives unsymmetrical dimers.[174]

Oxidation of anions to radicals that recombine to dimers is accomplished not only by electrolysis, but also by oxidants like I_2,[175] O_2,[176] or metal salts.[177]

b. Dehydrodimerization

Formation of alkyl radicals from hydrocarbons is possible via
hydrogen atom abstraction with initiators such as di-tert-butyl
peroxide, benzoyl peroxide, acetyl peroxide, Fenton's reagent,
or tert-butyl-hyponitrite. Hydrocarbons, alcohols, ethers,
aldehydes, ketones, acids, and their derivatives have been
used as hydrogen donors. Recombination reactions of the
intermediate radicals yield the dimers.[178]

$$RH \xrightarrow{X\cdot} R\cdot \longrightarrow R-R$$

$(CH_3)_2NCHO$	$\xrightarrow[H_2O_2]{Fe^{2+}}$	$\left[(CH_3)_2N\overset{\overset{\displaystyle O}{\|}}{C}\right]_2$	77%
CH_3CO_2H	$\xrightarrow{Ac_2O_2}$	$(HO_2CCH_2)_2$	50%
CH_3COCH_3	$\xrightarrow{Bz_2O_2}$	$(CH_3COCH_2)_2$	30%
	$\xrightarrow{DTBP}$		80%
	$\xrightarrow{DTBP}$		73%
$C_6H_5CH_3$	$\xrightarrow{Bu^t_2N_2O_2}$	$(C_6H_5CH_2)_2$	62%

Viehe[179] has introduced a new synthetic method involving the
trapping of radicals 1 with alkenes 186 to form capto-
dative[179,180] substituted radicals 187. These radicals are
too unreactive for further attack at alkene 186 and, therefore, dimerize to 188.

c = capto (electron-withdrawing) substituent

d = dative (electron-donating) substituent

Very effective in these syntheses is the combination of RS and CN substituents.[181]

Hydrogen abstraction can also occur with photolytically excited ketones and imines. Thus, Koch[182] has shown that iminolactone **189** dimerizes to **191** via radical **190**, which is formed by hydrogen donation from isopropyl alcohol to the triplet state of **189**. At -15°C the DL:meso-ratio is 0.45.

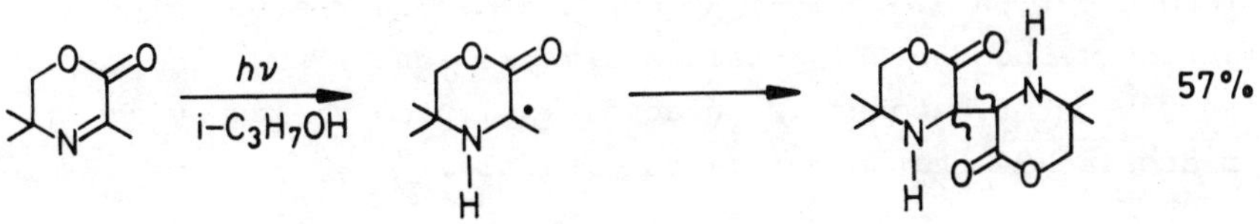

hν
i-C3H7OH
57%

189 **190** **191**

Elad[183] used photolytically excited ketones to abstract hydrogen from glycine, which combines with radicals to form new amino acids. In this way he synthesized peptide **193** from **192** via radicals **194** and **195** and observed a small optical induction of the C-C bond formation.

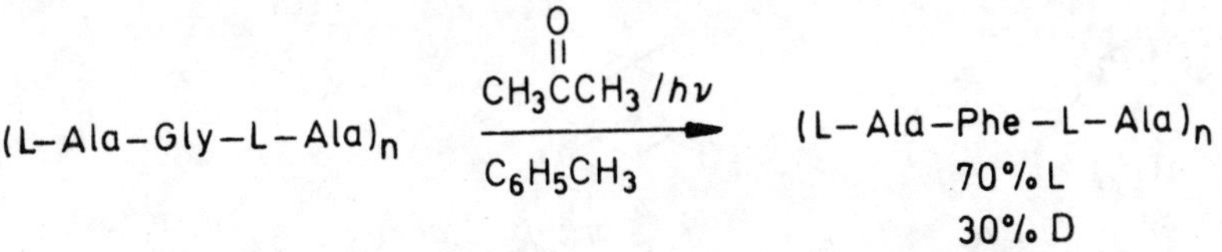

(L-Ala-Gly-L-Ala)n
CH3CCH3 /hν
C6H5CH3
(L-Ala-Phe-L-Ala)n
70% L
30% D

192 **193**

CH3-C-CH3
(L-Ala-NH-CH2-C-L-Ala)n
(L-Ala-NH-CH-C-L-Ala)n

194

CH3-C-CH3
CH3
CH2

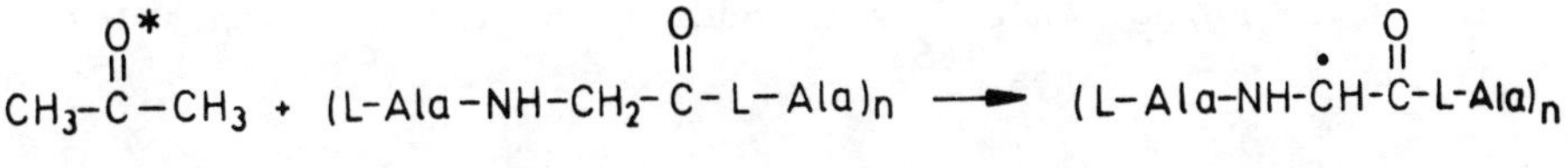

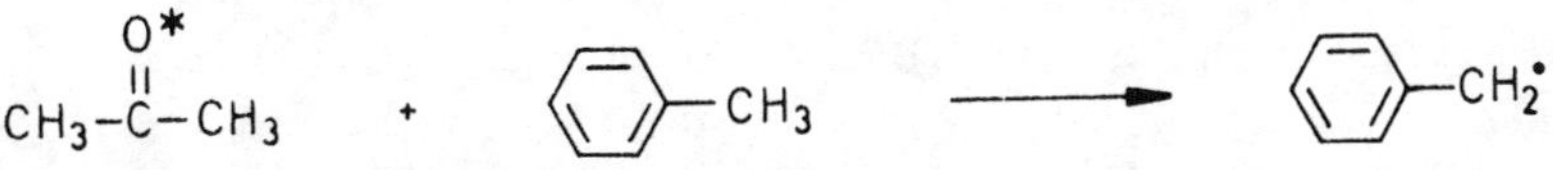

195

2. Reductive coupling

Reductive coupling of carbonyl compounds via radical-radical
recombination is mainly accomplished either electrolytically
at the cathode or by metals and metal ions. Thus, the cathodic
reduction of ketones in an acidic solution yields hydroxyalkyl
radicals **196** that dimerize to pinacols.[184]

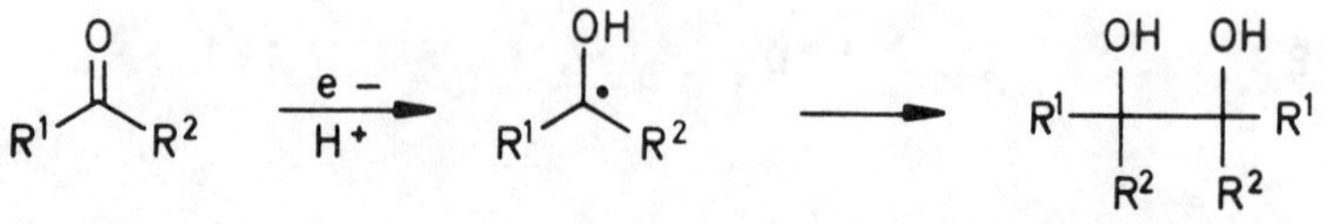

196

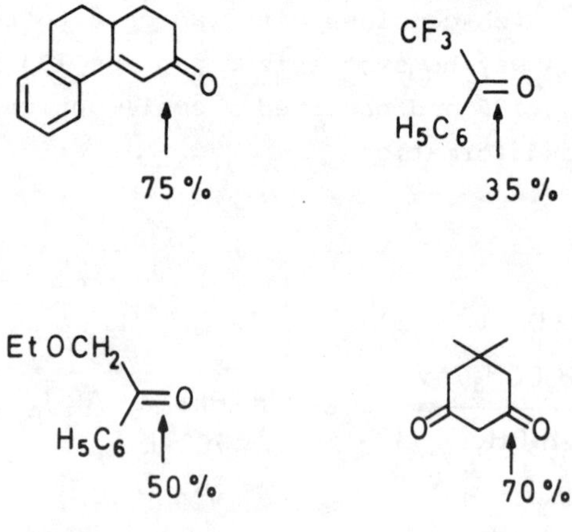

75 %

35 %

50 %

70 %

The pinacol dimerization can also be carried out with metals
like Mg or Mg/Hg. Under these conditions, ketyl radicals **197**
are the intermediates.[185]

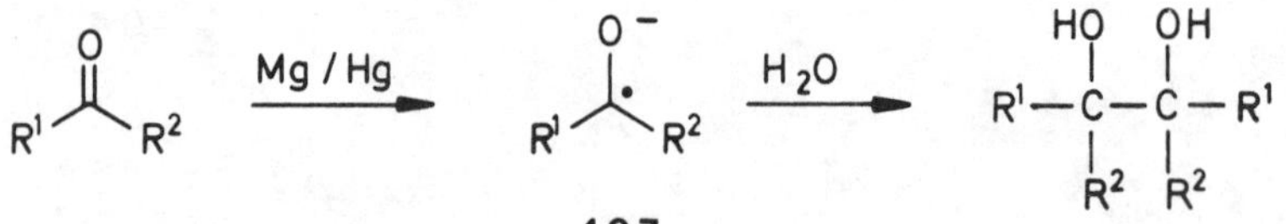

197

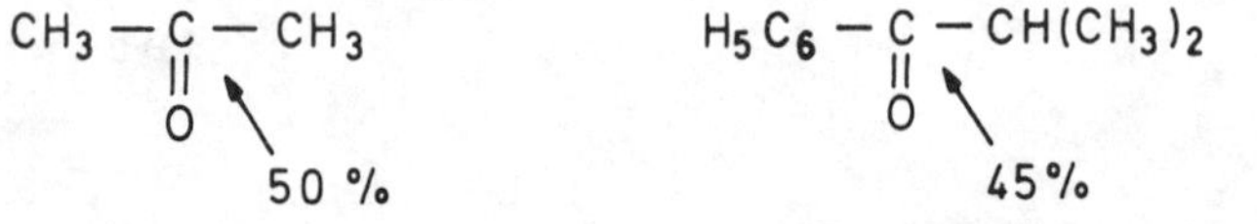

$CH_3 - C - CH_3$
$\quad\quad \overset{\|}{O}$

50 %

$H_5C_6 - C - CH(CH_3)_2$
$\quad\quad\quad \overset{\|}{O}$

45%

Ti(III) induced coupling of carbonyl compounds is believed to proceed via ketyl radical intermediates which dimerize to pinacol dianions to give alkenes after deoxygenation on the Ti(0) surface.[186]

Pinacols are formed by the photoreduction of aryl ketones in the presence of suitable hydrogen donors.[187] Ketones in the triplet state abstract a hydrogen atom from the alcohol and recombination of the newly formed radicals gives the pinacols. The synthesis of mixed recombination products is also possible.[188]

ROS-E*

In the acyloin reaction, electrons are transferred from Na to esters. The resulting radical salts **198** recombine to give α-diketones which are reduced to acyloins.[189]

Cathodic dimerization of electron-poor alkenes, e.g. the synthesis of adipodinitrile from acrylonitrile, which often occurs via anions and Michael addition,[118] will not be discussed here.

The reductive coupling of alkyl halides with metals, for example the Wurtz reaction, or with anions occurs in some cases via radicals generated by single electron transfer steps.[190] These radical C-C bond forming reactions make possible the formation of very strained compounds.[191]

In many cases the mechanism is not known in detail. Radical
and ionic reaction steps compete with each other and product
mixtures are formed.[190]

3. Diacylperoxides and azo compounds

Photolysis and thermolysis of initiators like diacylperoxides
or azo compounds produce radicals which, in the absence of
suitable traps, give combination and disproportionation
products. Synthetic applications of these reactions for intermolecular formation of C-C bonds are scarce because they have
few advantages over other reactions such as the Kolbe
electrolysis of acids. One advantage, however, is the greater
selectivity of the low temperature photolysis of mixed
diacylperoxides in the solid state.[192]

$$HO_2C(CH_2)_n \overset{O}{\overset{\|}{C}}-O-O-\overset{O}{\overset{\|}{C}} C_{11}H_{23} \quad \xrightarrow[\substack{-78°C \\ 65-70\%}]{h\nu} \quad HO_2C(CH_2)_n C_{11}H_{23}$$

Allyl radicals formed in the solid state have restricted
mobility and react with a much higher regioselectivity than
those formed under the conditions of the Kolbe electrolysis.
Even radicals with three different substituents at the radical
center react with high retention if chiral diacylperoxides
are used as precursors.[192]

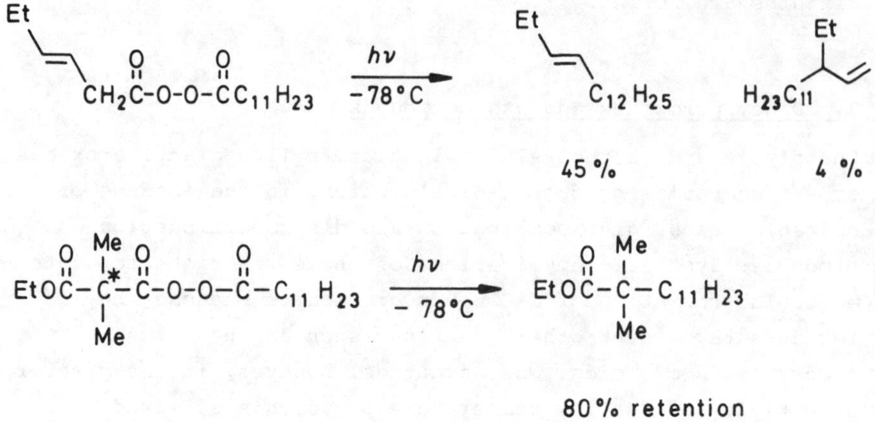

45 %   4 %

80 % retention

Azo compounds are mainly used in cyclization reactions, but Rüchardt[193] has shown that in cases where other synthetic routes because of steric effects fail, azo compounds still give C-C bond formation products.

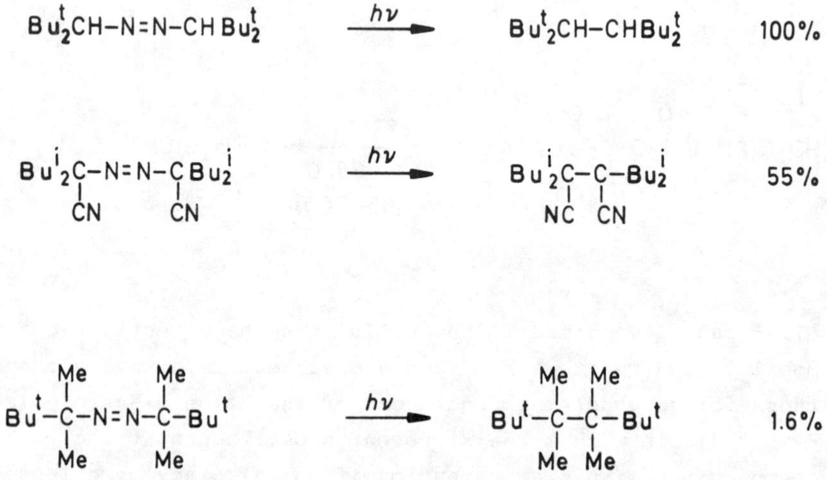

$Bu^t_2CH-N=N-CHBu^t_2$ $\xrightarrow{h\nu}$ $Bu^t_2CH-CHBu^t_2$ 100 %

$Bu^i_2C-N=N-CBu^i_2$ $\xrightarrow{h\nu}$ $Bu^i_2C-C-Bu^i_2$ 55 %

$\quad\ \ CN\qquad\ CN$ $NC\ \ CN$

80 % retention

The recombination of phenyl-substituted radicals with three different substituents often occurs with interesting diastereo-selectivity [193] which is mainly caused by entropy effects. Porter[194] has demonstrated that the diastereoselectivity of the recombination of radicals with polar substituents can be increased in lipid bilayers.

REFERENCES

1.   B. Giese, *Angew. Chem. Int. Ed. Engl.* **1983**, *22*, 771.

2.   G.M. Whitesides, J. San Filippo, *J. Am. Chem. Soc.* **1970**, *92*, 6611; C.L. Hill, G.M. Whitesides, *J. Am. Chem. Soc.* **1974**, *96*, 870.

3.   M. Devaud, *J. Organomet. Chem.* **1984**, *220*, C 27.

4.   B. Giese, G. Kretzschmar, *Chem. Ber.* **1984**, 3160.

5.   R.P. Quirk, R.E. Lea, *J. Am. Chem. Soc.* **1976**, *98*, 5973 ; G.A. Russell, D. Guo, *Tetrahedron Lett.* **1984**, *25*, 5239.

6.   B. Giese, *Angew. Chem. Int. Ed. Engl.* **1985**, *24*, 553.

7.   B. Giese, G. Kretzschmar, *Chem. Ber.* **1982**, *115*, 2012.

8.   B. Giese, J. Meister, *Chem. Ber.* **1977**, *110*, 2588.

9.   B. Giese, J.A. González-Gómez, T. Witzel, *Angew. Chem. Int. Ed. Engl.* **1984**, *23*, 69.

10. C. Chatgilialoglu, K.U. Ingold, J.C. Scaiano, *J. Am. Chem. Soc.*, **1981**, *103*, 7739.

11. L.H. Johnston, J. Lusztyk, D.D.M. Wagner, A.N. Abeywick-reyma, A.L.J. Beckwith, J.C. Scaiano, K.U. Ingold, *J. Am. Chem. Soc.* **1985**, *107*, 4594.

12. P. Pike, S. Hershberger, J. Hershberger, *Tetrahedron Lett.* **1985**, *26*, 6289.

13. H. Straub, K.P. Zeller, H. Leditschke in Houben-Weyl: *Methoden der Organischen Chemie*, Vol. 13/2b, Thieme, Stuttgart **1974**.

14. B. Giese, G. Kretzschmar, *Chem. Ber.* **1984**, *117*, 3175.

15. B. Giese, J. Meixner, *Tetrahedron Lett.* **1977**, 2783.

16. B. Giese, S. Lachhein, *Angew. Chem. Int. Ed. Engl.* **1982**, *21*, 768.

17. B. Giese, G. Kretzschmar, *Angew. Chem. Int. Ed. Engl.* **1981**, *20*, 965.

18. B. Giese, K. Heuck, *Chem. Ber.* **1979**, *112*, 3759; B. Giese, K. Heuck, *Tetrahedron Lett.* **1980**, *21*, 1829.

19. J. Barluenga, J. Lopez-Prado, P.J. Campos, G. Asensio, *Tetrahedron* **1983**, *39*, 2863; J. Barluenga, L. Ferrera; C. Nájera, M. Yus, *Synthesis* **1984**, 831.

20. B. Giese, T. Haßkerl, U. Lüning, *Chem. Ber.* **1984**, *117*, 859.

21. A.P. Kozikowski, T.R. Nieduzak, J. Scripko, *Organometallics* **1982**, 1, *675*.

22. R. Henning, H. Urbach, *Tetrahedron Lett.* **1983**, *24*, 5343.

23. B. Giese, K. Gröninger, *Tetrahedron Lett.* **1984**, *25*, 2743.

24. B. Giese, K. Heuck, *Chem. Ber.* **1981**, *114*, 1572.

25. S. Danishefsky, E. Taniyama, R.R. Webb, *Tetrahedron Lett.* **1983**, *24*, 11.

26. R.V. Stevens, K.F. Albizati, *J. Org. Chem.* **1985**, *50*, 632.

27. B. Giese, D. Bartmann, *Tetrahedron Lett.* **1985**, *26*, 1197.

28. B. Giese, D. Bartmann, unpublished results.

29. B. Giese, W. Zwick, *Chem. Ber.* **1979**, *112*, 3766; B. Giese, H. Horler, W. Zwick, *Tetrahedron Lett.* **1982**, *23*, 931.

30. B. Giese, W. Zwick, *Chem. Ber.* **1982**, *115*, 2526.

31. B. Giese, W. Zwick, *Tetrahedron Lett.* **1980**, *21*, 3569.

32. B. Giese, W. Zwick, *Chem. Ber.* **1983**, *116*, 1264; B. Giese, W. Zwick, unpublished results.

33. B. Giese, H. Horler, *Tetrahedron* **1985**, *41*, 4025.

34. D. Hoppe, *Angew. Chem. Int. Ed. Engl.* **1983**, *23*, 932.

35. B. Giese, H. Horler, *Tetrahedron Lett.* **1983**, *24*, 3221.

36. B. Giese, H. Horler, unpublished results.

37. H.O. House, L.J. Czuba, M. Gall, H.D. Olmstead, *J. Org. Chem.* **1969**, *34*, 2324.

38. D.A. Lindsay, J. Lusztyk, K.U. Ingold, *J. Am. Chem. Soc.* **1984**, *106*, 7087.

39. S. Wallowitz, J. Halpern, *J. Am. Chem. Soc.* **1984**, *106*, 8319.

40. B. Giese, U. Erfort, *Angew. Chem. Int. Ed. Engl.* **1982**, *21*, 130.

41. B. Giese, U. Erfort, *Chem. Ber.* **1983**, *116*, 1240.

42. B. Giese, R. Engelbrecht, U. Erfort, *Chem. Ber.* **1985**, *118*, 1289.

43. H.G. Kuivila, L.W. Menapace, C.R. Warner, *J. Am. Chem. Soc.* **1962**, *84*, 3584.

44. S.D. Burke, W.F. Fobare, D.M. Armistead, *J. Org. Chem.* **1982**, *47*, 3348.

45. B. Giese, J. Dupuis, *Angew. Chem. Int. Ed. Engl.* **1983**, *22*, 622.

46. J. Dupuis, B. Giese, D. Rüegge, H. Fischer, H.G. Korth, R. Sustmann, *Angew. Chem. Int. Ed. Engl.* **1984**, *23*, 896.

47. J. Dupuis, B. Giese, J. Hartung, M. Leising, H.G. Korth, R. Sustmann, *J. Am. Chem. Soc.* **1985**, *107*, 4332.

48. N. Ono, H. Miyake, A. Kamimura, I. Hamamoto, R. Tamura, A. Kaji, *Tetrahedron* **1985**, *41*, 4013.

49. B. Giese, D.B. Gerth, *J. Org. Chem.* **1986,** in press.

50. B. Giese, R. Rupaner, unpublished results.

51. R.M. Adlington, J.E. Baldwin, A. Basak, R.P. Kozyrod, *J. Chem. Soc. Chem. Commun.* **1983**, 944.

52. B. Giese, J. Dupuis, M. Leising, M. Nix, H.J. Lindner, *Carbohydr. Res.* **1986,** in press.

53. H.G. Korth, R. Sustmann, J. Dupuis, B. Giese, J. Chem.
     Soc. Perkin Trans 2 **1986**, in press.

54. B. Giese, T. Witzel, Angew. Chem. Int. Ed. Engl. **1986**,
     in press.

55. B. Giese, K. Gröninger, Tetrahedron Lett. **1984**, 25, 2743.

56. B. Giese, K. Gröninger, unpublished results.

57. B. Giese, H. Horler, M. Leising, Chem. Ber. **1986**, 119,
     444.

58. B. Giese, T. Witzel, unpublished results.

59. D.H.R. Barton, S.W. McCombi, J. Chem. Soc. Perkin Trans.1
     1975, 1574.

60. D.H.R. Barton, W.B. Motherwell in B.M. Trost, C.R. Hut-
     chinson (Eds): Organic Synthesis Today and Tomorrow,
     Pergamon, Oxford 1981.

61. N. Ono, H. Miyake, R. Tamura, A. Kaji, Tetrahedron Lett.
     1981, 22, 1705; N. Ono, R. Tamura, R. Tanikaga, A. Kaji,
     J. Chem. Soc. Chem. Commun. **1981**, 71.

62. D.D. Tanner, E.V. Blackburn, G.E. Diaz, J. Am. Chem. Soc.
     1981, 103, 1557; G.A. Russell, D.F. Dedolph, J. Org. Chem.
     1985, 50, 2498.

63. A.G. Davies, J.A.A. Hawari, C. Gaffney, P.G. Harrison,
     J. Chem. Soc. Perkin Trans. 2 **1982**, 631.

64. J. Lusztyk, B. Maillard, D.A. Lindsay, K.U. Ingold,
     J. Am. Chem. Soc. **1983**, 105, 3578.

65. B. Giese, B.E. Rückert, T. Witzel, unpublished results.

66. C.Chatgilialoglu, J.C. Scaiano, K.U. Ingold, Organo-
     metallics **1982**, 1, 466.

67. F.R. Cruickshank, S.W. Benson, Int. J. Chem. Kinet. **1969**,
     1, 381; T.J. Burkey, A.L. Castelhano, D. Griller,
     F.P. Lossing, J. Am. Chem. Soc. **1983**, 105, 4701;
     H. Birkhofer, H.D. Beckhaus, C. Rüchardt, Tetrahedron
     Lett. **1983**, 24, 185.

68. D.F. McMillen, D.M. Golden, *Ann. Rev. Phys. Chem.* **1982**,
     33, 493; W. Barbe, H.D. Beckhaus, C. Rüchardt, *Chem. Ber.*
     1983, *116*, 1042.

69. K.W. Egger, A.T. Cocks, *Helv. Chim. Acta* **1973**, *56*, 1516.

70. C. Walling, E.S. Huyser, *Org. Reactions* **1963**, *13*, 91;
     D. Elad in O.L. Chapman: *Organic Photochemistry*, Vol. 2,
     Marcel Dekker, New York 1969, p. 168.

71. W.H. Urry, F.W. Stacey, E.S. Huyser, O.O. Juveland,
     J.Am. Chem. Soc. **1954**, *76*, 450.

72. G.O. Schenck, G. Koltzenburg, H. Grossmann, *Angew. Chem.*
     1957, *69*, 177.

73. B. Fraser-Reid, N.L. Holder, D.R. Hicks, D.L. Walker,
     Can. J. Chem. **1977**, *55*, 3978.

74. R.L. Jacobs, G.G. Ecke, *J. Org. Chem.* **1963**, *28*, 3036.

75. B. Fraser-Reid, R.C. Anderson, D.R. Hicks, D.L. Walker,
     Can. J. Chem. **1977**, *55*, 3986.

76. T.J. Wallace, R.J. Gritter, *J. Org. Chem.* **1962**, *27*, 3067.

77. F. Minsci, *Top. Curr. Chem.* **1976**, *62*, 1.

78. P. Gottschalk, D.C. Neckers, *J. Org. Chem.* **1985**, *50*, 3498.

79. M.S. Kharasch, W.H. Urry, B.M. Kuderna, *J. Org. Chem.*
     1949, *14*, 248; T.M. Patrick, *J. Org. Chem.* **1952**, *17*,1009.

80. W.H. Urry, E.S. Huyser, *J. Am. Chem. Soc.* **1953**, *75*, 4876;
     L. Friedman, H. Shechter, *Tetrahedron Lett.* **1961**, 238.

81. G.I. Nikishin, G.V. Somov, A.D. Petrov, *Izvest. Akad.*
     Nauk **1961**, 2065; G.I. Nikishin, G.V. Somov, A.D. Petrov,
     Dokl. Akad. Nauk **1961**, *136*, 1099.

82. M. Hájek, J. Málek, *Synthesis*, **1976**,*8*, 315.

83. M.G. Vinogradov, S.P. Verenchikov, G.I. Nikishin,
     J. Org. Chem. USSR **1972**, 8, 2515.

84. A.D. Petrov, G.I. Nikishin, Y.N. Ogibin, *Dokl. Akad.*
     Nauk **1960**, *131*, 580.

85. G.I. Nikishin, Y.N. Ogibin, A.D. Petrov, *Dokl. Akad. Nauk*
     1961, *136*, 1326; J.C. Allen, J.I.G. Cadogan, B.W. Harris,
     D.H. Hey, *J. Chem. Soc.* **1962**, 4468; M. Cazaux, R. Lalande,
     Bull. Soc. Chim. Fr. **1971**, 2, 461.

86. D. Elad, R.D. Youssefyeh, *J. Chem. Soc. Chem. Commun.*
     1965, 7.

87. W.H. Urry, O.O. Juveland, *J. Am. Chem. Soc.* **1958**, *80*,
     3323.

88. J. Sinnreich, D. Elad, *Tetrahedron* **1968**, *24*, 4509.

89. F. Minisci, *Acc. Chem. Res.* **1975**, 165.

90. J. Metzger, J. Hartmanns, P. Köll, *Tetrahedron Lett.* **1981**,
     22, 1891.

91. M.S. Kharasch, E.V. Jensen, W.H. Urry, *Science* **1945**, *102*,
     128.

92. G. Sosnovsky, *Free Radicals in Preparative Organic
     Chemistry*, Macmillan, London 1964.

93. M.M. Martin, G.J. Gleicher, *J. Am. Chem. Soc.* **1964**, *86*,
     233; J.M. Tedder, *Angew. Chem. Int. Ed. Engl.* **1982**, *21*,
     401.

94. P. Boldt, L. Schulz, J. Etzemüller, *Chem. Ber.* **1967**, *100*,
     1281; M.H. Treder, H. Kratzin, H. Lübbecke, C.Y. Yang,
     P. Boldt, *J. Chem. Res. (S)* **1977**, 165.

95. D.M. Oldroyd, G.S. Fisher, L.A. Goldblatt, *J. Am. Chem.
     Soc.* **1950**, *72*, 2407; G. Dupont, R. Dulou, G. Clement,
     Bull. Soc. Chim. Fr. **1950**, 1056.

96. D.I. Davies in: *Essays on Free-Radical Chemistry*,
     Special Publication No. 24 of the Chemical Society,
     Burlington House, London 1970, p. 201.

97. B. Giese, K. Jay, *Chem. Ber.* **1977**, *110*, 1364.

98. D. Arlt, M. Jautelat, R. Lantzsch. *Angew. Chem. Int. Ed.
     Engl.* **1981**, *20*, 703.

99. P. Martin, E. Steiner, D. Bellus, *Helv. Chim. Acta* **1980**,
     63, 1947.

100. P. Martin, E. Steiner, J. Streith, T. Winkler, D. Bellus,
      Tetrahedron **1985**, *41*, 4057.

101. G.A. Kraus, K. Landgrebe, *Tetrahedron* **1985**, *41*, 4039.

102. D.H.R. Barton, D. Crich, G. Kretzschmar, *Tetrahedron*
      Lett. **1984**, *25*, 1055.

103. D.H.R. Barton, D. Crich, G. Kretzschmar, *J. Chem. Soc.*
      Perkin Trans. 1 **1986**, 39.

104. D.H.R. Barton, H. Togo, S.Z. Zard, *Tetrahedron* **1985**, *41*,
      5507.

105. D.H.R. Barton, H. Togo, S.Z. Zard, *Tetrahedron Lett.* **1985**,
      26, 6349.

106. D.H.R. Barton, D. Crich, *Tetrahedron Lett.* **1985**, *26*, 757.

107. H.C. Brown, M.M. Midland, *Angew. Chem. Int. Ed. Engl.*
      1972, 11, 692.

108. H.C. Brown, E. Negishi, *J. Am. Chem. Soc.* **1971**, *93*, 3777;
      H.C. Brown, G.W. Kabalka, W.M. Rathke, M.M. Rogič,
      J. Am. Chem. Soc. **1968**, *90*, 4165.

109. H.C. Brown, M.W. Rathke, G.W. Kabalka, M.M. Rogič,
      J. Am. Chem. Soc. **1968**, *90*, 4165.

110. H. C. Brown, G.W. Kabalka, *J. Am. Chem. Soc.* **1970**, *92*,
      712.

111. J.K. Kochi, *Acc. Chem. Res.* **1974**, *7*, 351.

112. E.I.Heiba,R.M. Dessau, W.J. Koehl, *J. Am. Chem. Soc.*
      1968, *90*, 5905; E. Heiba, R.M. Dessau, P.G. Rodewald,
      J. Am. Chem. Soc. **1974**, *96*, 7977.

113. W.E. Fristad, J.R. Peterson, A.B. Ernst, *J. Org. Chem.*
      1985, *50*, 3143.

114. E. Heiba, R.M. Dessau, *J. Org. Chem.* **1974**, *39*, 3456.

115. E.J. Corey, M.C. Kang, *J. Am. Chem. Soc.* **1984**, *106*, 5384;
      E.J. Corey, A.W. Gross, *Tetrahedron Lett.* **1985**, *26*,
      4291; W.E. Fristad, S.S. Hershberger, *J. Org. Chem.* **1985**,
      50, 1026.

116. G.I. Nikishin, M.G. Vinogradov, T.M. Fedorova, J. Chem.
      Soc. Chem. Commun. **1973**, 693; J.Y. Lallemand, Tetrahedron
      Lett. **1975**, 1217; M.E. Kurz, L. Reif, T. Tantrarat,
      J. Org. Chem. **1983**, 48, 1373.

117. F. Minisci, R. Galli, M.Cecere, V. Malatesta, T. Caronna,
      Tetrahedron Lett. **1968**, 5609; F. Minisci, P. Zammori,
      R. Barnardi, M. Cecere, R. Galli, Tetrahedron **1970**, 26,
      4153; F. Minisci, Synthesis **1973**, 1.

118. H.J. Schäfer, Angew. Chem. Int. Ed. Engl. **1981**, 20, 911.

119. H.J. Schäfer, A. Al Azrak, Chem. Ber. **1972**, 105, 2398.

120. D. Veltwisch, K.D. Asmus, J. Chem. Soc. Perkin Trans. 2
      1982, 1143; G.A. Russell, D. Guo, Tetrahedron Lett. **1984**,
      25, 5239.

121. R.C. Kerber, G.W. Urry, N. Kornblum, J. Am. Chem. Soc.
      1965, 87, 4520; N. Kornblum, W.J. Kelly, M.M. Kestner,
      J. Org. Chem. **1985**, 50, 4720; N. Kornblum, Angew. Chem.
      Int. Ed. Engl. **1975**, 14, 734.

122. G.A. Russell, W.C. Danen, J. Am. Chem. Soc. **1966**, 88,
      5663.

123. J.K. Kim, J.F. Bunnett, J. Am. Chem. Soc. **1970**, 92,
      7463; R.A. Rossi, R.H. de Rossi, Aromatic Substitution
      by the $S_{RN}1$ Mechanism, ACS Monograph 178, Washington
      1983.

124. B. Aebischer, R. Meuwly, A. Vasella, Helv. Chim. Acta
      1984, 67, 2236.

125. G.A. Russell, J. Hershberger, K. Owens, J.Am. Chem.
      Soc. **1979**, 101 1312; G.A. Russell, J. Hershberger,
      J. Organomet. Chem. **1982**, 225, 43; G.A. Russell,
      R.K. Khanna, Tetrahedron **1985**, 41, 4133.

126. See also: G.A. Russell, F. Ros, J. Am. Chem. Soc. **1985**,
      107, 2506; G.A. Russell, D.F. Dedolph, J. Org. Chem.
      1985, 50, 2378.

127. G.E. Keck, J.B. Yates, J. Am. Chem. Soc. **1982**, 104,
      5829.

128.   G.E. Keck, E.J. Enholm, J.B. Yates, M.R. Wiley,
       Tetrahedron 1985, 41, 4079.

129.   R.R. Webb, S. Danishefsky, Tetrahedron Lett. 1983, 24,
       1357.

130.   G.A. Keck, D.F. Kachensky, E.J. Enholm, J. Org. Chem.
       1985, 50, 4317.

131.   B. Giese, K. Gröninger, R. Sustmann, H.G. Korth, un-
       published results.

132.   J.E. Baldwin, D.R. Kelly, C.B. Ziegler, J. Chem. Soc.
       Chem. Commun. 1984, 133.

133.   G.A. Russell, H. Tashtoush, P. Ngoviwatchai, J. Am. Chem.
       Soc. 1984, 106, 4622.

134.   T.T.Tsou, M. Loots, J. Halpern, J. Am. Chem. Soc. 1982,
       104, 623; H.B. Gjerde, J.H. Espenson, Organometallics
       1982, 1, 435.

135.   M. Veher, K.N.V. Duong, F. Gaudemer, A. Gaudemer,
       J. Organomet. Chem. 1980, 177, 231; A.Bury, M.D. Johnson,
       J. Chem. Soc. Chem. Commun. 1980, 498; M.D. Johnson,
       Acc. Chem. Res. 1983, 16, 343.

136.   A. Gaudemer, K. Nguyen-Van-Duong, N. Shahkarami,
       S.S. Achi, M. Frostin-Rio, D. Pujol, Tetrahedron 1985,
       41,4095.

137.   S.N. Lewis, J.J. Miller, S. Winstein, J. Org. Chem.1972,
       37, 1478.

138.   G.E. Keck, J.H. Byers, J. Org. Chem. 1985, 50, 5442.

139.   D.H.R. Barton, D. Crich, Tetrahedron Lett. 1984,25, 2787.

140.   T.T. Tidwell in W.A. Pryor: Organic Free Radicals,
       ACS Symp. Ser. 69, 1978, p. 102.

141.   B. Maillard, A. Kharrat, F. Rakotomanana, E. Montaudon,
       C. Gardrat, Tetrahedron 1985, 41, 4047.

142.   K.U. Ingold, B.P. Roberts: "Free Radical Substitution
       Reactions", Wiley, New York 1971; A.J. Bloodworth,
       A.G. Davies, I.M. Griffin, B. Muggleton, B.P. Roberts,

J. Am. Chem. Soc. **1974**, 96, 7599; N.A. Porter, J.R. Nixon,
J. Am. Chem. Soc. **1978**, 100 7116; E.J. Corey, G. Schmidt,
K. Shimoji, Tetrahedron Lett. **1983**, 24, 3169.

143. P.I. Abell in C.H. Bamford, C.F.H. Tipper (ed):
Comprehensive Chemical Kinetics, Vol. XVIII, Elsevier,
Amsterdam, 1976, p. 111.

144. H.M. Bartels, P. Boldt, Liebigs Ann. Chem. **1981**, 40;
A. Bury, C.J. Cooksey, T. Funabiki, B.D. Gupta, M.D.
Johnson, J. Chem. Soc. Perkin Trans. 2 **1979**, 1050.

145. P.G. Gassman, J.L. Smith, J. Org. Chem. **1983**, 48, 4438.

146. H. Knoll, Z. Chem. **1982**, 22, 245.

147. M. Oyama, J. Org. Chem. **1965**, 30, 2429; J. Kollar,
U.S. Patent 183537, **1980**.

148. W.G. Bentrude, K.R. Darnall, J. Am. Chem. Soc. **1968**, 90,
3588; F. Minisci, R. Galli, M. Cecere, V. Malatasta,
T. Caronna, Tetrahedron Lett. **1968**, 5609; A. Citterio,
L. Filippini, unpublished results.

149. C. Pac, H. Sakurai, K. Shima, Y. Ogata, Bull. Chem. Soc.
Jap. **1975**, 48, 277.

150. G. Stork, P.M. Sher, J.Am. Chem. Soc. **1986**, 108, 302.

151. D.M. Blum, B.P. Roberts, J. Chem. Soc. Perkin Trans. 2
1978, 1313.

152. J. Luchetti, A. Krief, Tetrahedron Lett. **1978**, 2697;
B. Dechamps, M.C. Roux-Schmitt, L. Wartski, Tetrahedron
Lett. **1979**, 1377; C.A. Brown, A. Yamaichi, J. Chem. Soc.
Chem. Commun. **1979**, 100; Y. Tamaru, T. Harada, H. Iwamoto,
Z. Yoshida, J. Am. Chem. Soc. **1978**, 100, 5221; M.R.Binns,
R.K. Haynes, T.L. Houston, R. Jackson, Tetrahedron Lett.
1980, 573; N. Seuron, L. Wartski, J. Seyden-Penne,
Tetrahedron Lett. **1981**, 22, 2175; J. Lucchetti, A. Krief
Tetrahedron Lett. **1981**, 22, 1623.

153. G. Stork, G.L. Nelsen, F. Rouessac, O. Gringore, J. Am.
Chem. Soc. **1971**, 93, 3091; J.E. McMurry, W.A. Andrus,
J.H. Musser, Synth. Commun. **1978**, 8, 53; M.P. Cooke,
Tetrahedron Lett. **1979**, 2199; S.H. Liu, J. Org. Chem.
1977, 42, 3209.

154.   J. Gilman, R.H. Kirby, *J. Am. Chem. Soc.* **1941**, *63*, 2046;
       T. Caronna, A. Citterio, A. Clerici, R. Galli, *Org.*
       Prep. Proced. Int. **1974**, *6*, 299; M. Isobe, S. Kondo,
       N. Nagasawa, T. Goto, *Chem.Lett.***1977**, 679; T. Shono,
       I. Nishiguchi, M. Sasaki, *J. Am. Chem. Soc.* **1978**,
       4314; C. Petrier, J.C. de Souza Barbosa, C. Dupuy, *100*,
       J.L. Luche, *J. Org. Chem.* **1985**, *50*, 5761.

155.   G.H. Posner: *An Introduction to Synthesis Using Organo-
       copper Reagents*, Wiley, New York 1980.

156.   R. Scheffold, M. Dike, S. Dike, T. Herold, L. Walder,
       J. Am. Chem. Soc. **1980**, *102*, 3642.

157.   R. Scheffold, *Chimia* **1985**, *39*, 203.

158.   S. Albrecht, R. Scheffold, *Chimia* **1985**, *39*, 211; A. Ruhe,
       L. Walder, R. Scheffold, *Helv. Chim. Acta* **1985**, *39*, 1301.

159.   R. Scheffold, R. Orlinski, *J. Am. Chem. Soc.* **1983**, *105*,
       7200.

160.   S.Torii: *Electroorganic Synthesis*, Verlag Chemie,
       Weinheim 1985.

161.   L. Eberson, *Acta Chem. Scand.* **1959**, *13*, 40; B.C.L. Wee-
       don, *Adv. Org. Chem.* **1963**, 1, 1; J. Haufe, F. Beck,
       Chem. Ing. Tech. **1970**, *42*, 170.

162.   F. Fichter, S. Lurie, *Helv. Chim. Acta* **1933**, *16*, 885.

163.   F.L. Pattison, J.B. Stothers, R.G. Woolford, *J.Am. Chem.*
       Soc. **1956**, *78*, 2255.

164.   J.P. Waefler, P. Tissot, *Electrochim. Acta* **1978**, *23*, 899.

165.   R.F. Garwood, N. ud Din, C.J. Scott, B.C.L. Weedon,
       J. Chem. Soc. Perkin Trans. 1 **1973**, 2714.

166.   E.J. Corey, R.R. Sauers, *J. Am. Chem. Soc.* **1959**, *81*,
       1739.

167.   G. Stork, A. Meisels, J.E. Davies, *J. Am. Chem. Soc.*
       1963, *85*, 3419.

168.   W. Seidel, J. Knolle, H.J. Schäfer, *Chem. Ber.* **1977**,
       110, 3544; J. Knolle, H.J. Schäfer, *Angew. Chem. Int.*
       Ed. Engl. **1975**, *14*, 758; H. Klünenberg, H.J. Schäfer,

Angew. Chem. Int. Ed. Engl. **1978**, *17*, 47.

169. T. Okubo, S. Tsutsumi, *Bull. Chem. Soc. Jpn.* **1964**, *37*,
      1794; R. Brettle, J.C. Parkin, *J. Chem. Soc. C.* **1967**,
      1352; M. Lacani, I. Tabakovic, M. Vukicevic, *Croat.
      Chem. Acta* **1973**, *45*, 465.

170. H.J. Schäfer, *Chem. Ing. Tech.* **1969**, *41*, 179.

171. R. Bauer, H. Wendt, *J. Electroanal. Chem.* **1977**, *80*, 395.

172. M.Y. Fioshin, L.A. Salmin, L.A. Mirkind, A.G. Kornienko,
      Zh. Obshch. Khim. **1965**, *10*, 594; H.J. Schäfer, *Angew.
      Chem. Int. Ed. Engl.* **1970**, *9*, 158; H.J. Schäfer,
      R. Pistorius, *Angew. Chem. Int. Ed. Engl.* **1972**, *11*, 48.

173. H.J. Schäfer, E. Steckhan, *Angew. Chem. Int. Ed. Engl.*
      1969, 8, 518; D. Koch, H.J. Schäfer, E. Steckhan,
      Chem. Ber. **1974**, *107*, 3640; see also: B. Belleau,
      Y.K. Au-Young, *Can. J. Chem.* **1969**, *47*, 2117; M.A. Le
      Moing, G. Le Guillanton, J. Simonet, *Electrochim. Acta*
      1981, *26*, 139.

174. J.R. Falck, L.L. Miller, F.R. Stermitz, *Tetrahedron*
      1974, *30*, 931.

175. R. Schlecker, U. Henkel, D. Seebach, *Chem. Ber.* **1977**,
      110, 2880.

176. G.A. Russell, G. Kaup, *J. Am. Chem. Soc.* **1969**, *91*, 3851;
      see also: G. Eglington, W. McCrae in R.A. Raphael,
      E.C. Taylor, H. Wynberg (ed): *Advances in Organic
      Chemistry: Methods and Results* Vol. IV, Intercience,
      London 1963, p. 225.

177. P. Yates, Y.C. Toong, *J. Chem. Soc. Chem. Commun.* **1978**,
      205.

178. For references see: J. Protasiewicz, G.D. Mendenhall,
      J. Org. Chem. **1985**, *50*, 3220.

179. H.G. Viehe, R. Merényi, L. Stella, Z. Janousek, *Angew.
      Chem. Int. Ed. Engl.* **1979**, *18*, 917.

180. H.G. Viehe, Z. Janousek, R. Merényi, *Acc. Chem. Res.*
      1985, *18*, 148.

181. S. Mignani, M. Beaujean, Z. Janousek, R. Merényi,
      H.G. Viehe, *Tetrahedron* (Woodward issue) **1981**, *37*, W 111.

182. R.W. Bennett, D.L. Wharry, T.H. Koch, *J. Am. Chem. Soc.*
      1980, *102*, 2345.

183. M. Schwarzberg, J. Sperling. D. Elad, *J. Am. Chem. Soc.*
      1973, *95*, 6418.

184. E. Kariv, E. Gileadi, *Collect. Czech. Chem. Commun.* **1971**,
      36, 476; C.P. Andrieux, J.M. Saveant, *Bull. Soc. Chim.*
      Fr. **1972**, 3281; C.P. Andrieux, J.M. Saveant, *Bull. Soc.*
      Chim. Fr. **1973**, 2090; E. Touboul, G. Dana, *J. Org. Chem.*
      1979, *44*, 1397.

185. H.O. House: *Modern Synthetic Methods*, Vol. II, Benjamin,
      Menlo Park 1972, p. 167.

186. J.E. McMurry, *Acc. Chem. Res.* **1983**, *16*, 405.

187. H.D. Becker, *J. Org. Chem.* **1967**, *32*, 2140.

188. J.C. Gramain, R. Remuson, Y. Troin, *J. Chem. Soc. Chem.*
      Commun. **1976**, 194.

189. J.J. Bloomfield, D.C. Owsley, J.M. Nelke, *Organic*
      Reactions **1976**, *23*, 259.

190. J.F. Garst, *Acc. Chem. Res.* **1971**, 4, 400; B.J. Wakefield:
      The Chemistry of Organolithium Compounds, Pergamon Press,
      New York 1974; W.C. Danen in E.S. Huyser (ed): *Methods*
      in Free Radical Chemistry, Vol. V, Dekker, New York 1974,
      p. 1; R.D. Guthrie in E. Buncel, T. Durst (ed): *Compre-*
      hensive Carbanion Chemistry, Elsevier, Amsterdam 1980,
      p. 230; see also: W.A. Bailey, R.P. Gagnier, J.J. Patri-
      cia, *J. Org. Chem.* **1984**, *49*, 2098; E.C. Ashby, T.N.Pham,
      B. Park, *Tetrahedron Lett.* **1985**, *26*, 4691.

191. M.A. Flamm-ter Meer, H.D. Beckhaus, K. Peters, H.G. von
      Schnering, C. Rüchardt, *Chem. Ber.* **1985**, *118*, 4665.

192. M. Feldhues, H.J. Schäfer, *Tetrahedron* **1985**, *41*, 4213.

193. R. Winiker, H.D. Beckhaus, C. Rüchardt, *Chem. Ber.*
      1980, *113*, 3456; W. Barbe, H.D. Beckhaus, H.J. Lindner,

C. Rüchardt, *Chem. Ber.* **1983**, *116*, 3235 ; C. Rüchardt,
H.D. Beckhaus, *Angew. Chem. Int. Ed. Engl.* **1985**, *24*,
529; see also: J.W. Timberlake, Y.M. Jun, *J. Org. Chem.*
1979, *44*, 4729; P.S. Engel, *Chem. Rev.* **1980**, *80*, 99.

194.   D.G. Wujek, N.A. Porter, *Tetrahedron* **1985**, *41*, 3973.
W.J. Brittain, N.A. Porter, P.J. Krebs, *J. Am. Chem.
Soc.* **1984**, *106*, 7652.

Chapter 4

Intramolecular Formation of Aliphatic C—C Bonds

A. Introduction

Radical cyclization reactions developed in the last years
represent a breakthrough for synthetic radical chemistry.[1]
These reactions exhibit interesting regioselectivities and
stereoselectivities and can be carried out with a variety of
functional groups as radical traps. Since their activation
entropies are less negative than those of intermolecular
reactions, not only carbon-carbon, but also carbon-oxygen and
carbon-nitrogen multiple bonds efficiently react intramolecularly with radicals. Thus, the entropy of cyclization of the
5-hexenyl-radical 1 is 14 e.u. larger (less netagive) than that
of the intermolecular addition of an ethyl radical to 1-hexene,
but the activation enthalpies are not much different from each
other.[2]

$$\Delta H^{\ddagger} = 5.5 \text{ kcal/mol}$$
$$\Delta S^{\ddagger} = -11 \text{ e.u.}$$

1   **2**   **3**

98 : 2

$$\overset{\displaystyle .}{\underset{C_4H_9}{\diagdown}} + \overset{\displaystyle .}{C_2H_5} \quad \xrightarrow[\Delta S^{\ddagger} = -25 \text{ e.u.}]{\Delta H^{\ddagger} = 5.6 \text{ kcal/mol}} \quad \underset{H_9C_4}{\overset{.}{\bigcirc}}$$

The cyclization of the 5-hexenyl radical **1** to cyclopentylmethyl radical **2** has a rate constant of about 10^6 (s^{-1}) at $20^{\circ}C$,[3] and is increased by electron-withdrawing substituents at the double bond.[4] Thus, cyclization reactions are fast enough for a successful application in syntheses.

Although radicals **1** and **2** have the same nucleophilicity, the selectivity requirement is also fulfilled because radical **1** reacts intramolecularly with the olefinic bond, whereas radical **2** reacts intermolecularly, e.g. with Bu_3SnH.

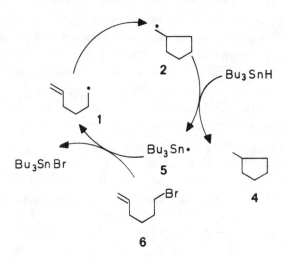

Since both the reactivity and the selectivity requirements for chain reactions are fulfilled, radical cyclization reactions are synthetically useful. But neither three- nor four-membered rings can be synthesized because the cyclo-propylmethyl and cyclobutylmethyl radicals **8** and **10** cleave

the rings with high rates to form the more stable acyclic
radicals **7** and **9**, respectively.[5]

7   8

9   10

Cyclized products can be formed only in systems where a large
proportion of the ring strain is already present in the
non-cyclized radical, such as the norbornenyl radical **11**,[6] or
if the cyclic radicals are stabilized by substituents, e.g.
the phenyl substituted radical **14**.[7]

11   12

13   14

In contrast to cyclopropylmethyl and cyclobutylmethyl radicals
8 and **10**, cyclopropyl and cyclobutyl radicals **15** and **16** undergo

ring cleavage very slowly,[8] but one cannot generate these radicals from allyl or homoallyl radical precursors.[5,9]

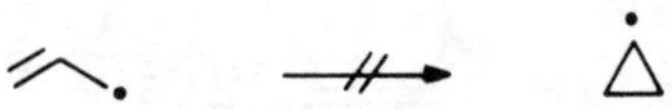

15

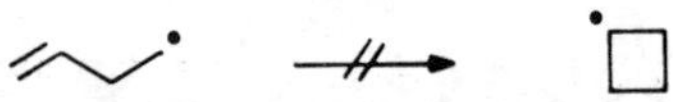

16

The formation of medium sized rings is also not a synthetically useful route,[10] but recently Porter[11] has shown that large rings can be synthesized via radical cyclization.

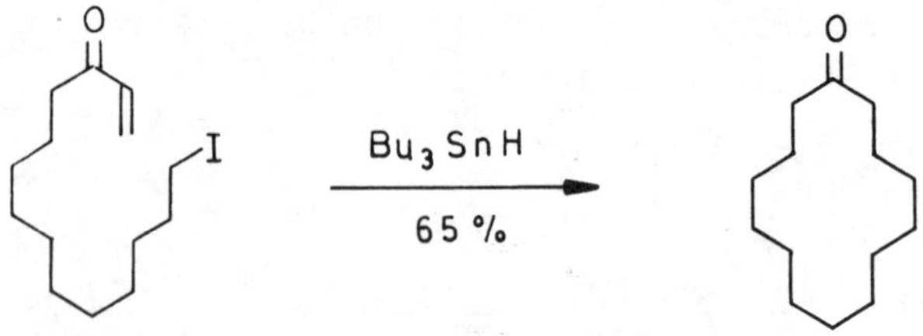

1. Regioselectivity

5-Hexenyl and 5-heptenyl radicals **1** and **17** cyclize predominately to the smaller rings **2** and **18**, respectively. Thus, the less stable prim. radicals are formed faster than the more stable sec. radicals.[10]

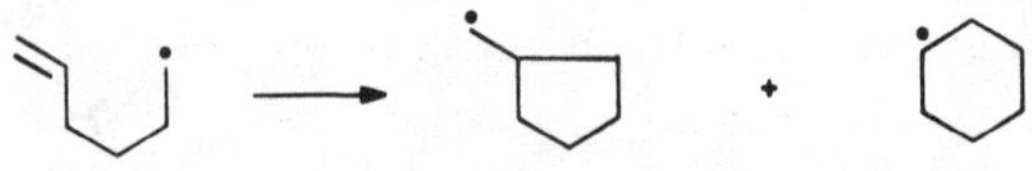

1   **2**   **3**

98   :   2

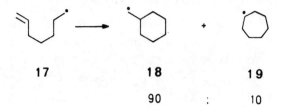

17	**18**	**19**
	90 :	10

According to the Baldwin rules,[12] the formation of **2** and **18** proceeds via a favored *exo*-cyclization, whereas **3** and **19** are formed in *endo*-cyclization reactions. Beckwith[10] has pointed out that stereoelectronic effects can explain these regio-selectivities. Since the addition of alkyl radicals to alkenes proceeds via the unsymmetrical transition state **20**.[13] the same arrangement of atoms in the intramolecular addition favors transition states **21** and **22**.[10]

R•
\
⸴⸴⸴C=C⸴⸴⸴
◣   ◥

20

21   **22**

Additional steric effects[14] resulting from a 1,3-diaxial interaction in the transition state, as well as less fa-vorable entropies[15] in the formation of cycloalkyl radicals compared to cycloalkyl methyl radicals, could also retard the *endo*-cyclization. Calculations using force field methods indicate that the smaller rings are formed because of less strained transition states.[10]

The ratio of the five- to six-membered rings depends on the substituents at the olefinic bond. A methyl group at the olefinic carbon atom reduces the attack at that C-atom due to steric effects; however, methyl groups at the radical center have only small effects.[10]

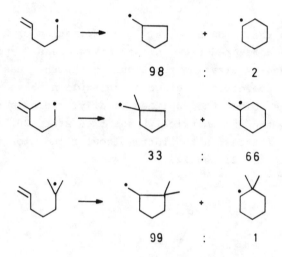

The formation of five-membered rings also predominates when CC-triple bonds are used as intramolecular radical traps or if an oxygen atom is within the chain.[10,16]

The situation changes, however, if the radical cyclization is reversible. Julia[17] has shown that under thermodynamic conditions six-membered rings are predominately formed.

A silicon group in the chain can also favor six-membered ring formation, as was observed by Wilt.[18] In a similar manner, systems containing amido groups [19,20] often yield to a considerable amount the larger ring.

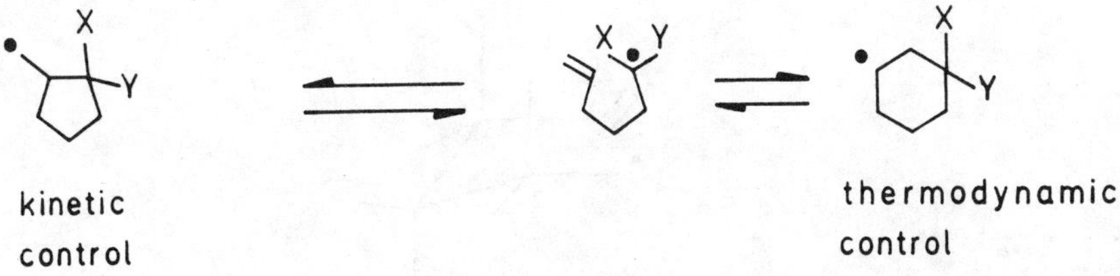

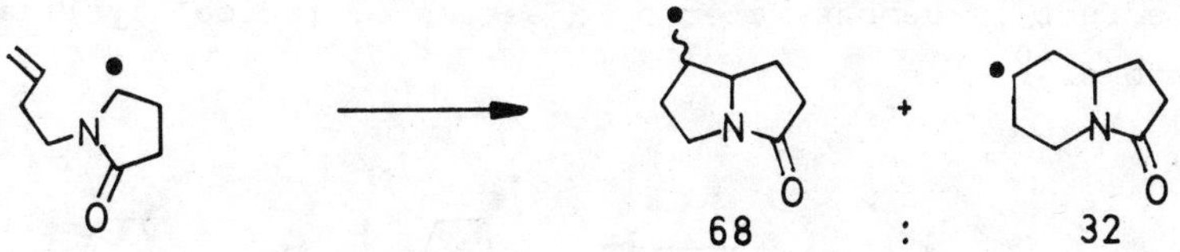

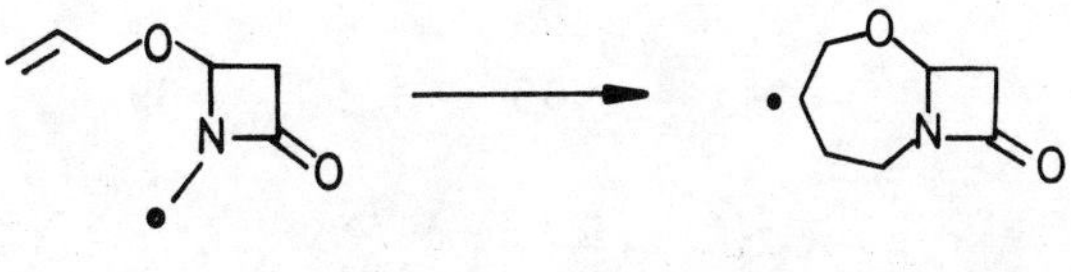

kinetic
control

thermodynamic
control

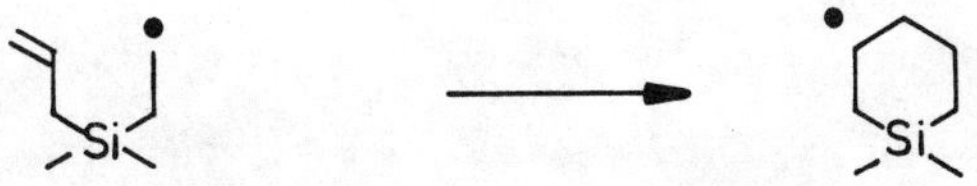

68   :   32

2. Stereoselectivity

Two useful guidelines governing the ring closure of substituted hexenyl radicals were formulated by Beckwith:[21]

- 1- or 3-substituted radicals preferentially give *cis*-disubstituted cyclopentyl products;

-   2- or 4-substituted radicals give mainly *trans*-disubsti-
    tuted cyclopentyl derivatives.

These rules can be explained by the transition state structure
23, in which axial and equatorial positions are clearly
distinguishable at C-2, C-3, and C-4. Thus, the more favorable
conformer should contain the substituents in the equatorial
position.

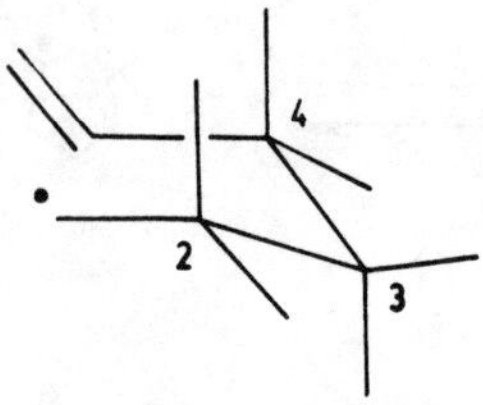

23

This effect is demonstrated by a series of radical cyclization
reactions.[10]

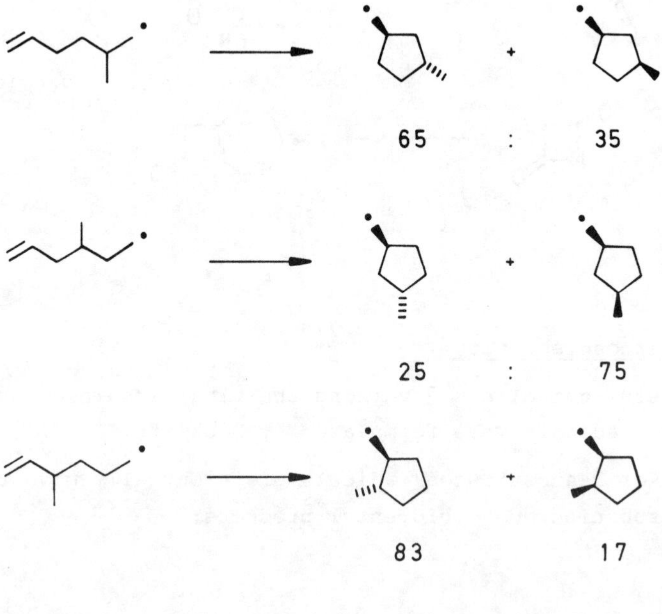

65   :   35

25   :   75

83   17

The preferred formation of the cis-cyclopentyl system from radical **24** cannot be so easily explained.[10] One possibility is that in the early transition states of radical additions to alkenes, in which the distance between the reacting C-atoms is about 2.3 Å,[13] attractive interactions between the methyl group and the π-bond exist. This explanation is backed by the observation of Corey that **25** cyclizes to the cis-isomer **26**.[22]

24   **27**   :   **73**

25   **26**

Porter[23] has shown that the ratio of cyclized radicals **28:29** depends on the solvent mixture. In the presence of micelles or bilayer forming tensides, the amount of cis-isomer **28** increases. This effect was explained by a special conformation of radical **27** at the water/benzene interphase.

27   **28**   **29**

C_6H_6 / H_2O	50	:	50
C_6H_6 / H_2O / tenside	70	:	30

Interestingly, in the formation of a six-membered ring from the 1-methylheptenyl radical **30** the *cis*-product is again the main isomer.[10]

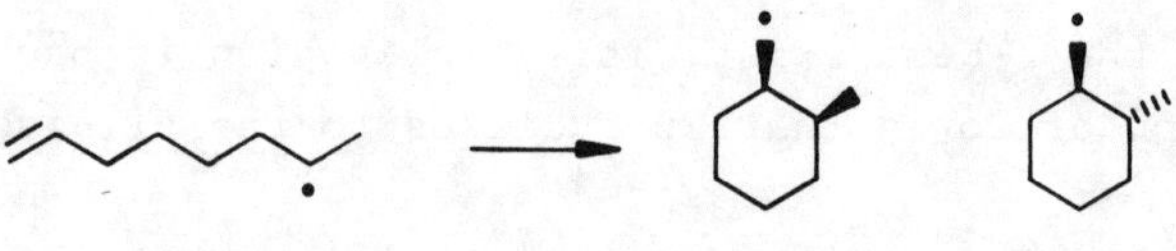

30   73   :   27

In radical cyclizations of cyclic alkenes *cis*-junction is the main reaction.[12]

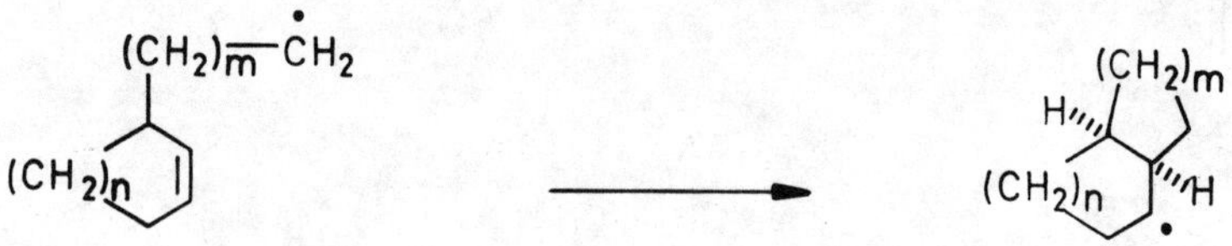

n = 1,2 ; m = 2,3

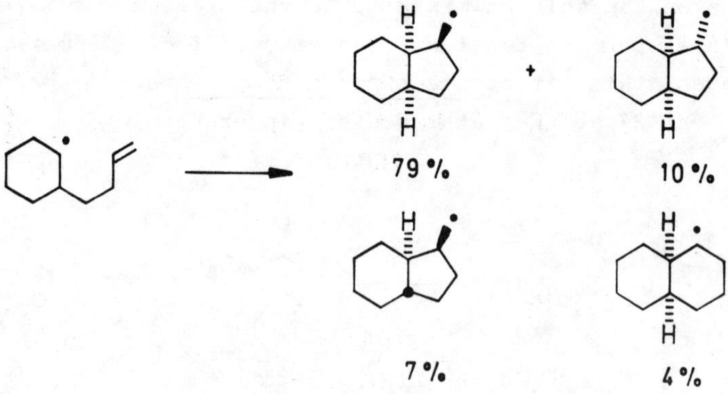

79 %   10 %

7 %   4 %

B. Trapping with Hydrogen Donors

1. Mercury hydride

As described on page 38, the reduction of alkylmercury salts with inorganic hydrides like $NaBH_4$ or Bu_3SnH gives alkyl radicals via mercury hydrides and alkylmercury radicals.

$$RHgX \xrightarrow{NaBH_4} RHgH \xrightarrow[-HX]{+X\cdot} RHg\cdot \xrightarrow{-Hg} R\cdot$$

The flexibility of this method lies in the easy generation of alkylmercury salts, for example, by mercuration of alkenes. Thus, Danishefsky synthesized cyclic products via solvomercuration of dienes.[24]

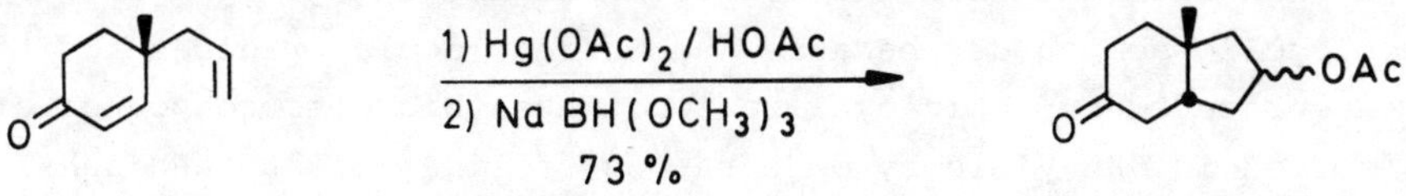

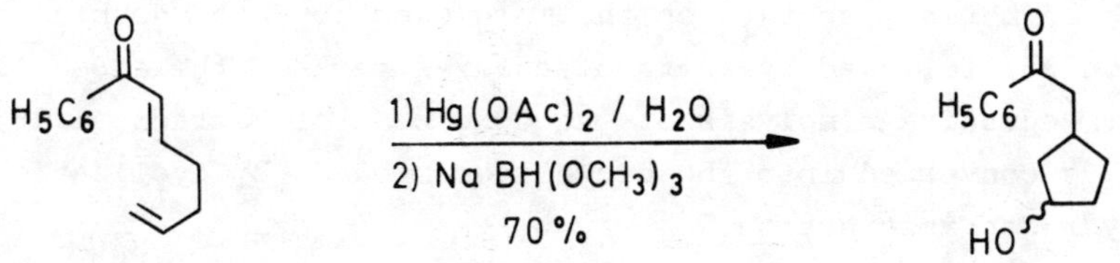

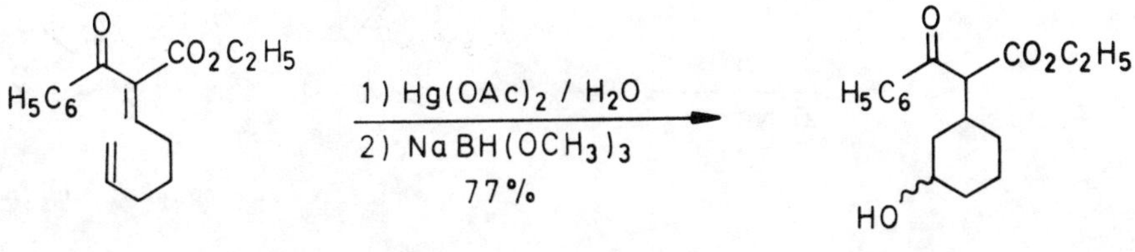

1) $Hg(OAc)_2$ / $HOAc$
2) $NaBH(OCH_3)_3$
73 %

1) $Hg(OAc)_2$ / H_2O
2) $NaBH(OCH_3)_3$
70 %

1) $Hg(OAc)_2$ / H_2O
2) $NaBH(OCH_3)_3$
77 %

Organomercury salts **31**,which are formed as intermediates, are reduced without isolation to give radical **33**. After intra-molecular addition to the CC-double bond, the cyclized radicals **34** are trapped by the organomercury hydride **32**, yielding products **35**.

31   **32**

33   **34**   **35**

Dienes with suitable nucleophilic neighboring groups give bicyclic compounds because the intramolecular radical addition is preceded by ionic cyclization. Thus, bicyclic compounds are formed in 68% yield by mercuration of linalool **36**, and sub-sequent reduction. The reaction proceeds via the monocyclic organomercury compound **37**, which is formed by electrophilic attack of mercury acetate on the more electron-rich double bond in **36**, followed by intramolecular attack of the alcohol group. Reductive homolysis of the C-Hg bond gives radical **38**, which is converted into the bicyclic compounds by cyclization and hydrogen abstraction.[25]

36   1) Hg(OAc)$_2$ 2) NaBH$_4$ 52%   +   16%

37   **38**

An amide function can also act as a nucleophile for the ionic reaction step; thus, benzannulated bicyclic compounds are formed from diene **39**.[24]

39   75   :   25

Corey has used this methodology in a prostaglandin synthesis.[22] Starting from the ester of arachidonic acid, the hydroperoxide **40** is synthesized, which, after an ionic mercuration reaction, yields the cyclic product **41**. Radical **25** (page 149), formed by reduction with Bu$_3$SnH, cyclizes to **26** which is trapped by O$_2$ and the mercury hydride to give **42**. The synthesis is completed by deoxygenation with (C$_6$H$_5$)$_3$P and yields the cis-isomer **43**.

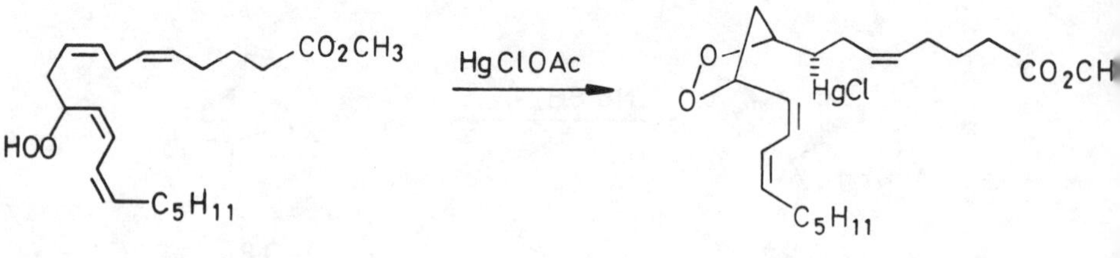

40 → HgClOAc → **41**

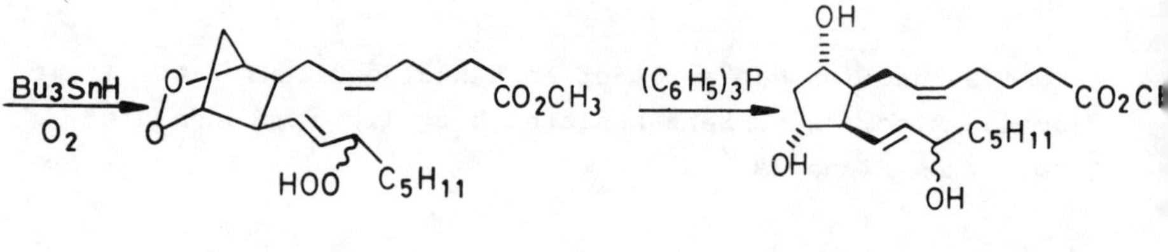

42 → **43**

2. Tin hydride

Before 1980, studies of radical cyclizations centered mainly
on the mechanistic aspects of the reaction.[1,5,17,18] Early
synthetic applications include the synthesis of tepenoid **44** by
Bakuzis and dihydroagarofuran **45** by Büchi.[26]

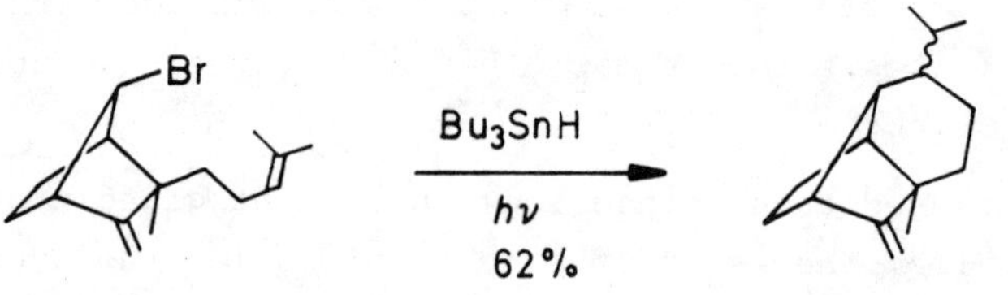

Bu$_3$SnH

hν

62%

44

Bu$_3$SnH

Δ/AIBN

67%

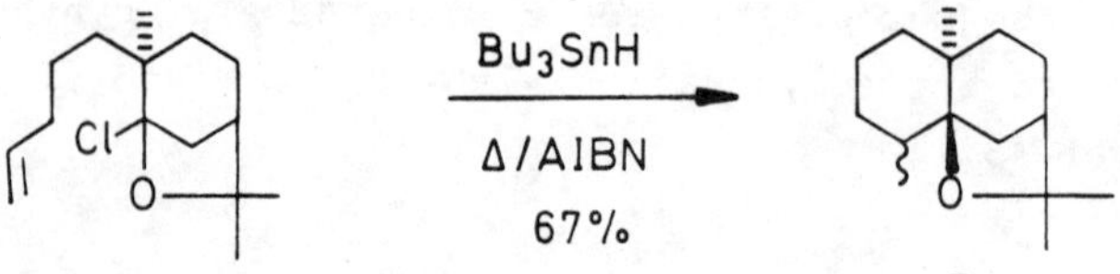

45

In these reactions, halides and tin radicals react to give
alkyl radicals which, after cyclization, abstract a hydrogen
atom from Bu₃SnH (see p. 56). In the last few years this
reaction has been applied to syntheses of target molecules
in several laboratories. Stork[27] has carried out pioneering
work in which both alkyl and vinyl radicals were used in
cyclization reactions with both acyclic and cyclic double- and
triple-bonds. The synthesis of the butenolide **49** is an elegant
example.[28]

The radical chain reaction takes place in boiling benzene with
bromide **47**, which is easily obtained from **46**. The tributyltin
radical abstracts the bromine atom from **47**, leaving the less
reactive C-Cl bond intact. The thus formed alkyl radical **50**
attacks the triple bond in an *exo*-cyclization forming the
vinyl radical **51**, which then reacts with the double bond of
the cyclohexene ring. Radical **52** intramolecularly reacts with
tributyltin hydride to give **48** and the tributyltin radical,
which once again cleaves the C-Br bond of **47**.

During the synthesis, the radical center is transferred from a prim. alkyl to a vinyl, and then to a cyclohexyl position in two consecutive cyclizations. These reactions occur with very high rates, high yields, and a high stereoselectivity in the annulation step.

Other examples of such tandem cyclizations have been worked out by Curran[29] for the synthesis of cyclopentanoid natural products in which three cyclopentanes are annulated. The decisive radical cyclization in the synthesis of hirsutene 55 begins with iodide 54, which is synthesized from lactone 53.

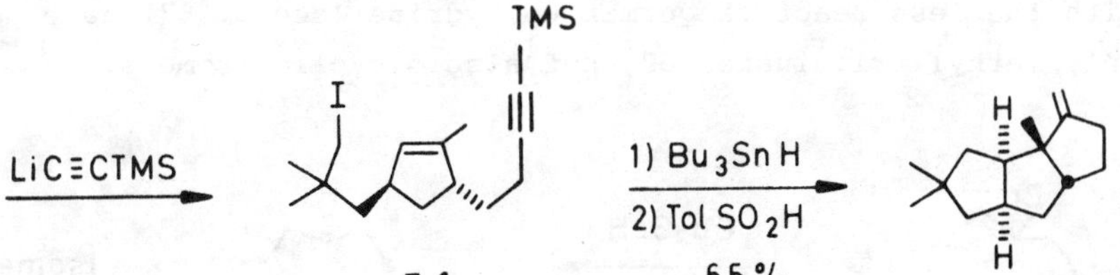

54

1) Bu$_3$SnH
2) Tol SO$_2$H
65%

55

In the synthesis of capnellene **58**, lactone **56** is transformed into the bromide **57**; radical tandem cyclization with Bu$_3$SnH then gives capnellene **58**.[30]

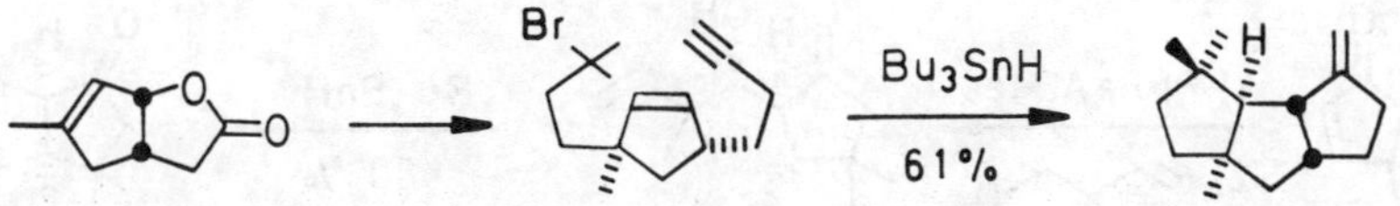

56   **57**   **58**

Bu$_3$SnH
61%

Radical precursor for the synthesis of silphiperfol-6-ene is bromide **60**, which is generated from **59** in three steps. Radical cyclization yields **61**, a mixture of silphiperfol-6-ene and 9-episilphiperfol-6-ene.[31]

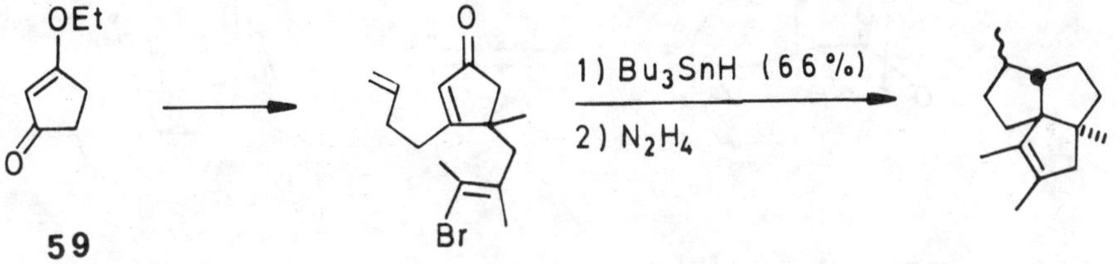

59   **60**   **61**

1) Bu$_3$SnH (66%)
2) N$_2$H$_4$

Beckwith[32] started with the acyclic precursor 62. Cyclization with the less reactive germanium hydride (see p. 67) gave not only methyl triquinanes 63, but also bicyclic isomers.

A synthesis of bicyclic compound 64 has been carried out by Livinghouse.[33]

Hart[34] synthesized lactone 66 as a key intermediate for pleurotine 65a and geogenine 65b.

66

65a : X = ß - H
65b : X = α - H

Aldehyde **67**, which can be made from benzoic acid, is converted to **68**. Cyclization of **68** gives tricyclus **69** which yields **70** in six steps.

The predominant formation of **69** shows that the C-C and the C-H bonds are formed with high stereoselectivity. Because of the particular conformation of the bicyclic radical **71**, the C-C bond formation leads to a *trans*-annulation of the new ring. Remarkable also is the high diastereoselectivity of the hydrogen abstraction from the acyclic alkyl radical. The conformation **72**, in which the carbonyl groups of the lactone and the ester are far apart from each other, seems to be the preferred conformation of this acyclic radical. Hydrogen transfer from the least hindered side then gives the major isomer **69**.

Another application of radical cyclization reactions, which was developed by Hart,[35] is for the formation of pyrrolizidine alkaloids. The synthesis of (-)-dehydrohastanecine 75 starts with malic acid, which is converted to the radical precursor 73. Cyclization gives 74 and after two additional steps, alkaloid 75.

The cyclized compound **74** is also a precursor for (+)-heliotri-
dine **76** and (+)-hastanecine **77**.[35]

74 ⟶

76

74 ⟶

77

For the syntheses of alkaloids **75–77**, it is important that the
alkyne **73** is substituted with a terminal Me_3Si-group   because
less bulky substituents lead to a considerable amount of
six-membered ring formation.[35]

$$R = H \qquad\qquad < 5 \quad : \quad > 95$$
$$n - C_3H_7 \qquad 56 \quad : \quad 44$$
$$i - C_3H_7 \qquad 65 \quad : \quad 35$$
$$t - C_4H_9 \quad > 95 \quad : \quad < 5$$
$$SiMe_3 \quad > 95 \quad : \quad < 5$$

The formation of the larger ring was also observed in annulation reactions of ß-lactams carried out by Bachi.[36] But again, the size of the ring depends on the substitution. With a terminal triple bond, a seven-membered ring is formed, whereas a phenylated alkyne yields a six-membered ring.

The larger amount of *endo*-cyclization is not caused by the triple bond because alkenes show a similar behavior. It is, therefore, the effect of amido substitution that leads to the successful competition of the formation of larger rings.[20,37]

R = H	< 5	:	> 95
CO_2Me	68	:	4
C_6H_5	> 95	:	< 5

R¹	R²	R³		
H	H	H	68 :	32
Me	Me	H	> 95 :	< 5
H	H	Me	< 5 :	> 95

Speckamp[38] has shown that *endo*-cyclization is also favored with sulfonamide **78**. However, the byproduct **79** may result from *exo*-cyclization, followed by fragmentation.

Pyrrolidins, for example **80**, were synthesized by Padwa[39] in high yields.

80

Cyclization reactions starting from carbohydrates have also been carried out. Thus, Wilcox[40] synthesized the cyclopentanoid compounds **83** from the pentose **81** which leads to the radical precursor **82** in four steps. Cyclization gave the five-membered rings in 85% yield.

The *trans*-isomer **83a** is formed predominately, especially from the Z-isomer of **82**. This shows that **84a** is favored over **84b**.

84a   **84b**

The synthesis of butenolide **49** by Stork[28] has already shown that not only alkyl, but also vinyl radicals can lead to cyclized products (see p. 155).

Another example is the synthesis of norseychellanone **86** by cyclization of bromide **85**.[41]

85   **86**

Additional work was carried out by Marinovic,[42] who also showed that the yield increases in going from chloride via bromide to iodide.

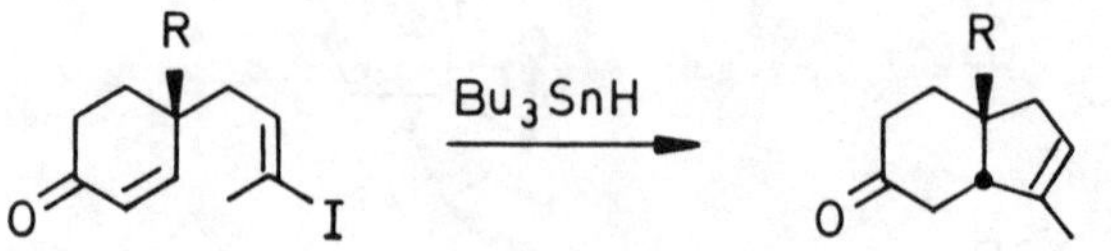

$$R = H \quad (85\%)$$
$$CH_3 \quad (83\%)$$

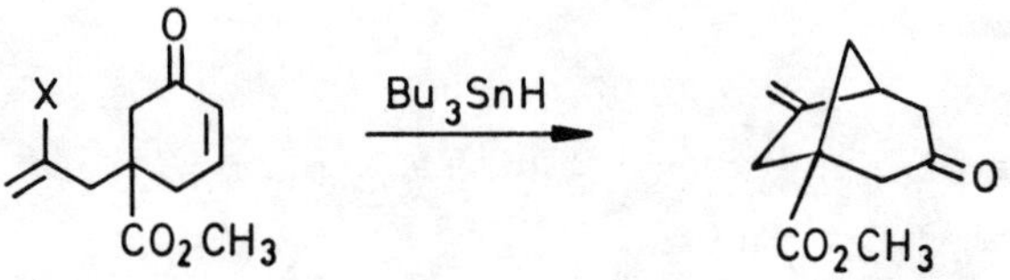

$$X = Cl \quad (59\%)$$
$$Br \quad (91\%)$$
$$I \quad (93\%)$$

The tin method has turned out to be a suitable method for the synthesis of γ-lactones. Three different approaches have been worked out by Stork,[43] Clive,[44] and Bachi.[45]

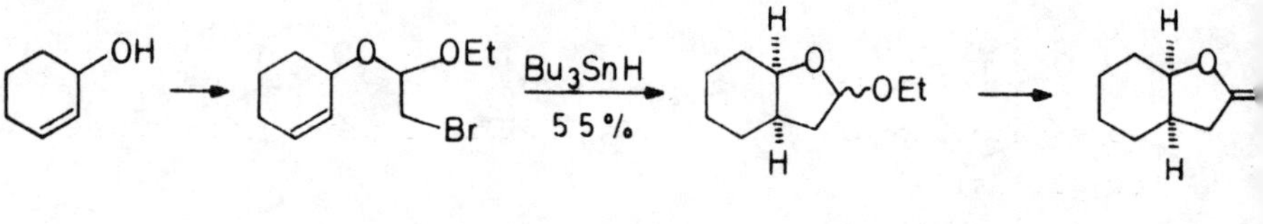

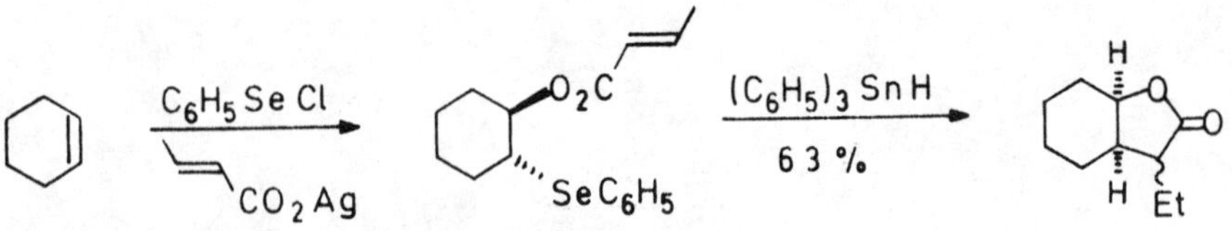

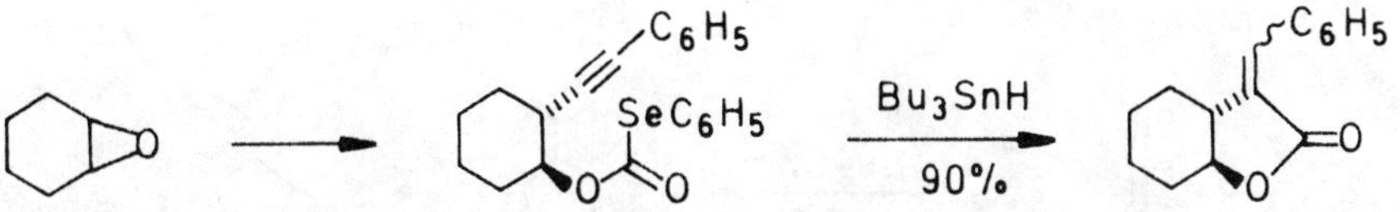

Pattenden[46] has shown that in these intramolecular reactions nucleophilic radicals can even add to the electron-rich double bond of an enolether.

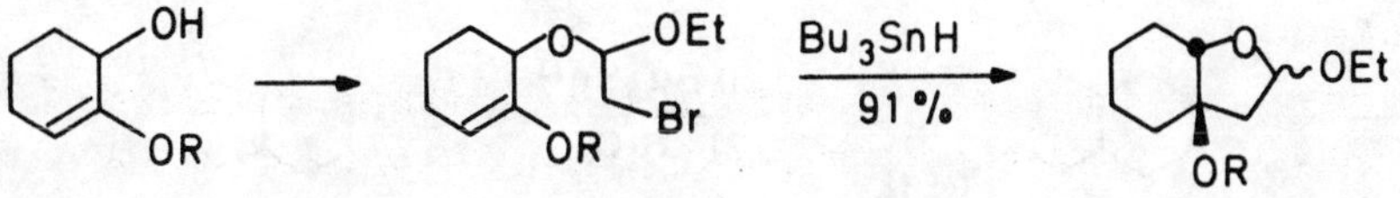

87

88

In the synthesis of **88**, the enolether function in **87** acts as a ketone equivalent. Ketones themselves are not good radical traps because the π-CO bond is too strong. Nevertheless, there are some examples of this kind in which products are formed in low yields.[47]

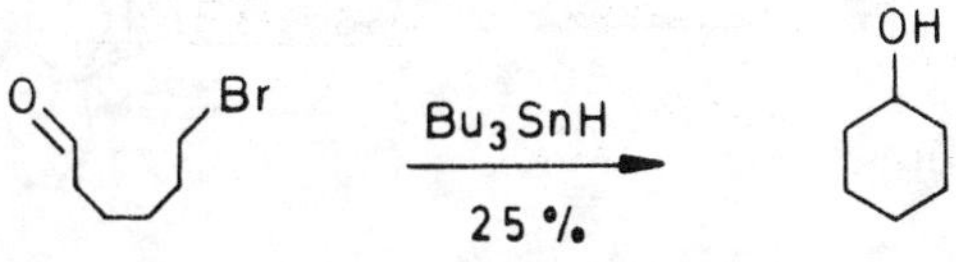

In contrast to ketones, allenes[48] and nitriles[49,50] are
suitable radical traps for cyclization reactions.

In Clive's[50] synthesis of ketone **90**, the carbon-centered radical
is generated from **89** by attack of the tin radical on the sulfur
atom.
Kraus[51] has demonstrated that radical precursors with a
bridgehead C-Br bond can be also used.

Nitro compounds are appropriate precursors for radical C-C bond forming reactions and Ono[52] has applied this methodology for cyclization reactions.

The often faster rate of cyclization compared to intermolecular C-C bond formation was used by Stork[53] to carry out both steps in one reaction sequence.

3. "Carbon hydride"

Examples for the trapping of cyclized radicals with C-H bonds as hydrogen donors are found in the earliest radical cyclization reactions performed by Julia[54] and his group. Hexenyl systems 91 give rise to cyclic products 92 if the radical chain is started by initiators which abstract a hydrogen atom from the tert. carbon atom. Since the cyclization is reversible with ester and nitrile substituted radicals, the more stable cyclohexyl radicals predominate yielding products 92.

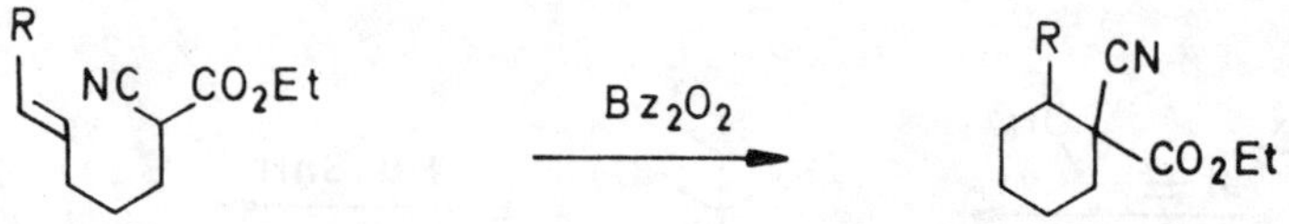

91

92

R = H (51%), CH$_3$(65%), C$_2$H$_5$(78%)

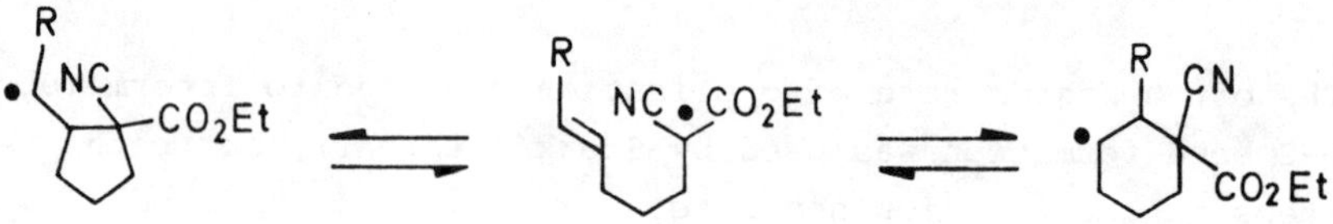

But alkyne 93 gives mainly the cyclopentyl derivative 94.[55]

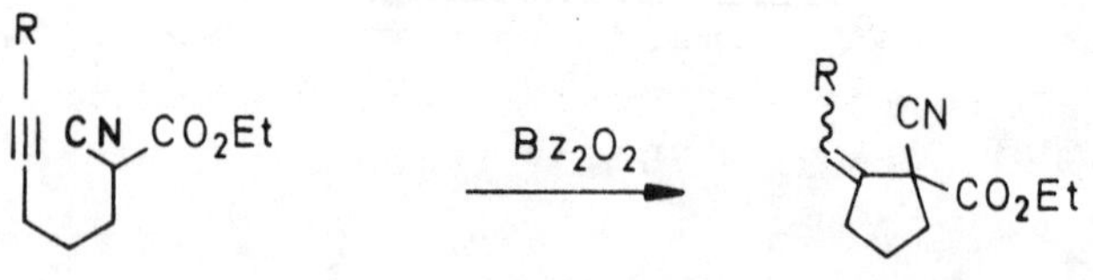

93

94

R = H (27%), CH$_3$(63%)

This procedure can also be applied to tandem cyclization reactions.[56]

Hydrogen abstraction can also occur from aldehydes or their derivatives. Thus, citronellal gives a mixture of menthone (50%) and isomenthone (25%).[57]

Barton[58] has photolytically cyclized **95** to the tetracycline derivative **96**.

95   **96**

Alcohols don't effectively activate C-H bonds and give low yields of cyclization products, e.g. the cyclization of citronellol to isomeric menthols.[59]

A different approach starts with dienes **97** which react with carbon centered radicals to give **98**. Cyclization yields radicals **99** which abstract a hydrogen atom from the added reagent RH.

97   **98**   **99**

Formamide, esters, lactones, aldehydes, and chloroform can be used as the hydrogen donor RH.[5,60,61]

C. Trapping with Heteroatom Donors

1. Halogen donor

The addition of halogenated alkanes to dienes with suitably oriented double bonds yields cyclic products whose amount depends on the rate of halogen abstraction. The electrophilic alkyl radicals attack one of the double bonds, cyclization occurs, and finally, halogen abstraction gives the product and the chain carrying electrophilic radical (see p. 77).[5,60,62]

100

Brace[63] has reported that in the reaction of diene **100**, the *trans*-isomer is the major product. Thus, radical **101** shows a different stereoselectivity than the radical which is generated from halide **102** (see p. 149).

Another exceptional example is the formation of the four-membered ring from the thermally induced addition of penta-fluoroiodoethane to diene **103**, observed by Modena.[64]

In cyclization reactions of dienes, carbon-free halogen donors can be used. Thus, Traynham[65] has shown that cyclodecadiene 104 gives 105 via radical chlorination.

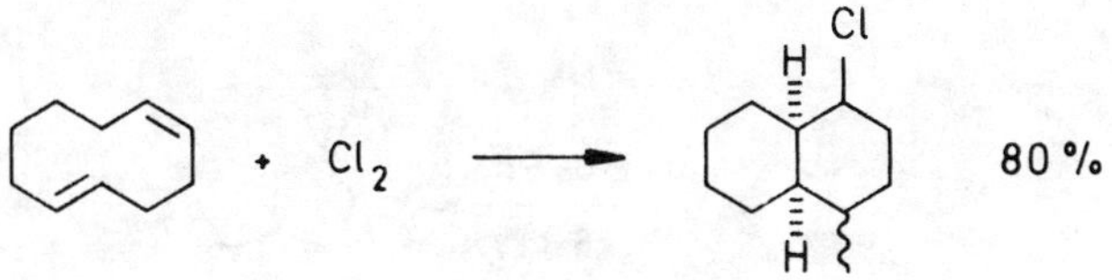

104

105

The alkene can also act as the halogen donor. This is, for example, the case if the halogen abstraction is facilitated by Cu-salts. Ogasawara[66] and Ito[67] have used this method for the synthesis of cyclopropane 106 and mesembrane 107, respectively.

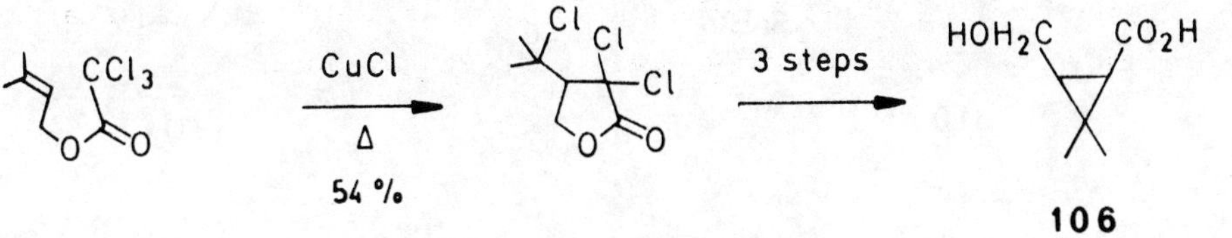

106

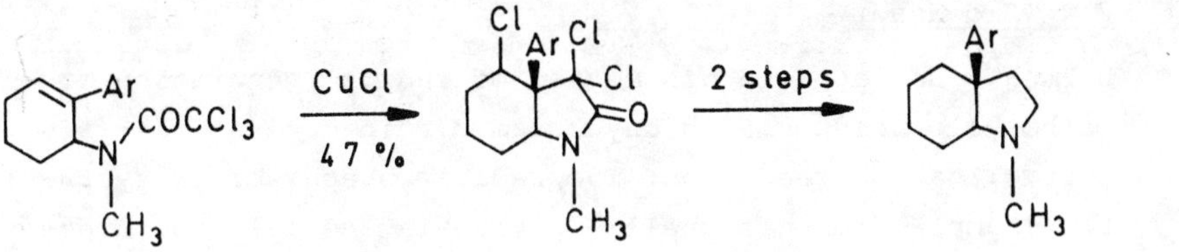

107

Weak tert. C-I bonds offer reaction possibilities in which the halogen donor is the molecule that cyclizes. Thus, Curran[68] has shown that iodide **108** yields **109** in the presence of radical initiators.

108   **109**

The synthesis of lactone **111** from iodide **110**, carried out by Kraus,[69] presumably occurs via iodine abstraction by the cyclized radical (see p. 81).

110   **111**

2. Thio donor

Barton[60] has applied his method of radical generation from carboxylic acids via thiohydroxamic acid derivatives (see p.82) to cyclization reactions. The radical precursor is synthesized, for example from citronellic acid, via the acid chloride and reaction with thiothiazole. Irradiation yields the cyclopentyl system **112**, which can be desulfurated to cedrene.

112

Starting with disulfides, cyclized radicals are trapped by thiyl radicals in a non-chain reaction. Thus, Kuehne[71] irradiated α-acoradiene **113** with dimethyl disulfide and obtained a nearly quantitative yield of **114**. Under similar conditions, α-bulnese **115** gives a product mixture containing **116**, which was desulfurated to dihydropatchulene.[71]

113   **114**

115   **116**

3. Oxygen donor

In the biosynthesis of prostaglandins molecular oxygen presumably traps the cyclized radical **117**.[72]

117

Porter[72] and Corey[22] have carried out syntheses that are similar to these in vivo reactions (see p. 154).

4. Metal donor

In the reaction of 2-allyloxycyclohexyl bromide **118** with cobaloxime (see p. 102), Tada[73] observed the formation of the cyclized organocobaloxime **119**.

This reaction proceeds via radical **120**, which is formed by electron transfer from cobaloxime to bromide **118**. Radical cyclization and trapping of **121** with cobaloxime yields reaction product **119**.

Organocobaloximes of this kind can undergo 1,2-elimination reactions. Thus, Pattenden[74] cyclized bromide **122** to **123**.

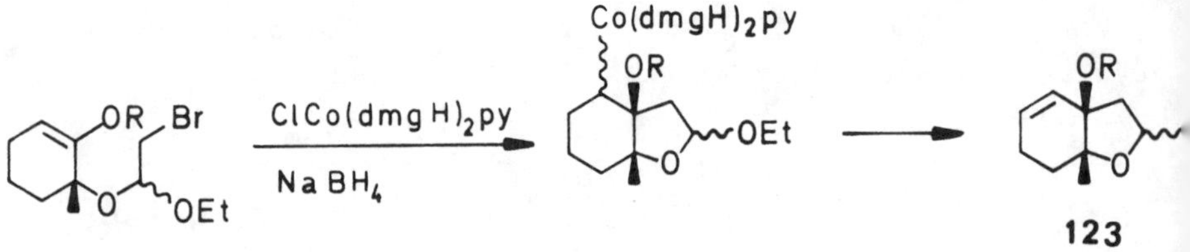

122 → **123** 71 %

The cobaloxime is generated in these reactions from chloro-cobaloxime with $NaBH_4$, but electrochemical reduction is also possible, as Torii[75] has shown.

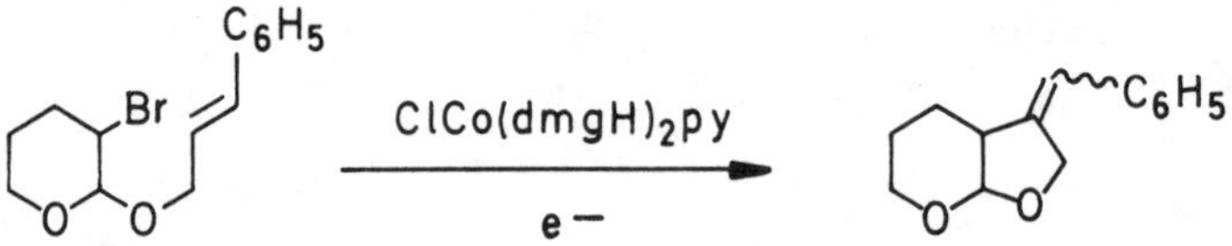

81%

With C-C triple bonds, 1,2-elimination cannot occur and olefins are formed after the cyclization step under reductive conditions.[76]

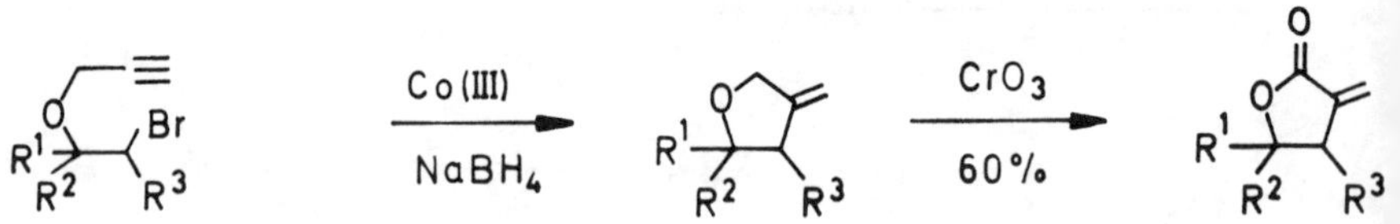

65 - 85%

65 - 75 %.

Scheffold[77] used vitamin B_{12} (see p. 111) in analogous reductive cyclization reactions.

These reactions are somewhat similar to the photolysis of hexynylcobaloxime **124** carried out by Johnson.[78] The reaction occurs via radical cyclization **126→127**.

D. Trapping by Electron Transfer Reactions

As in intermolecular C-C bond forming reactions, radicals can also be trapped after cyclization by electron transfer.

1. Oxidation

Corey,[79] Fristad,[80] and Snider[81] have used the oxidative addition of carboxylic acids to alkenes (see p. 90) for radical cyclization under mild conditions.

The cyclized radicals are oxidized to cations and either react with nucleophiles or are deprotonated to alkenes.

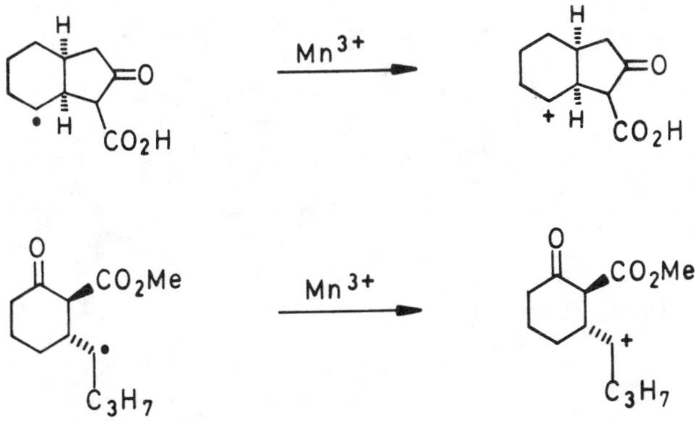

This reaction has been applied to the synthesis of podocarbic ester **129** by cyclization of **128** and Clemmensen reduction.[81]

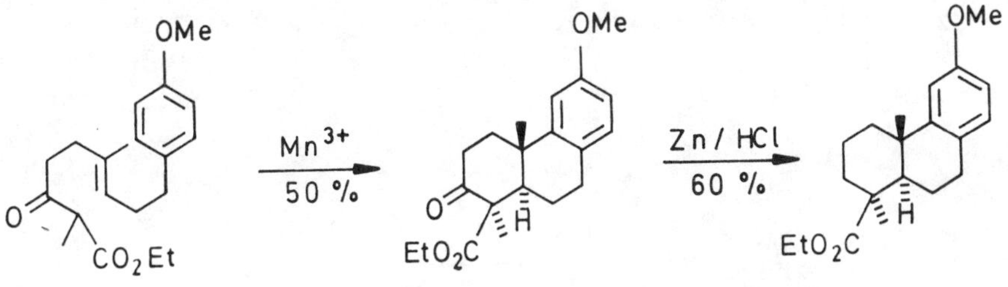

128

129

Benzoyloxy radicals, which are generated in a dibenzoyl peroxide/Cu$^+$ pair, add to one of the double bonds of dienes. Cyclization, followed by oxidation with Cu^{2+}, and finally deprotonation gives the product. Using this methodology, Breslow[82] transformed geranyl acetate to **130**, presumably via radical intermediates.

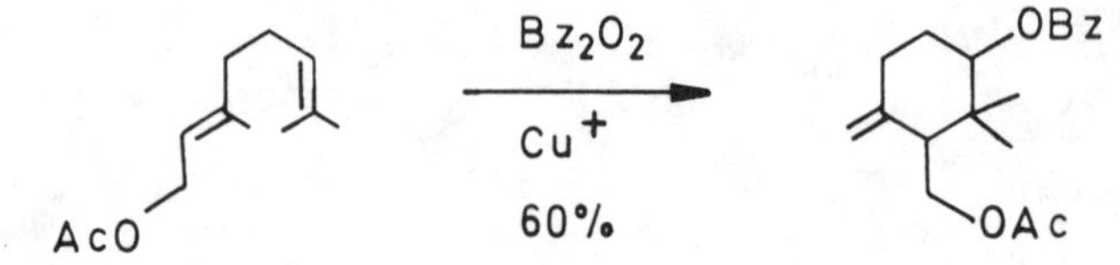

130

Kolbe electrolysis of unsaturated acids has been used by Schäfer[83] for the synthesis of tetrahydrofurans **131**. The products are formed via combination of the cyclized radical **132** with $R^3\cdot$, generated from an excess of acid R^3CO_2H.

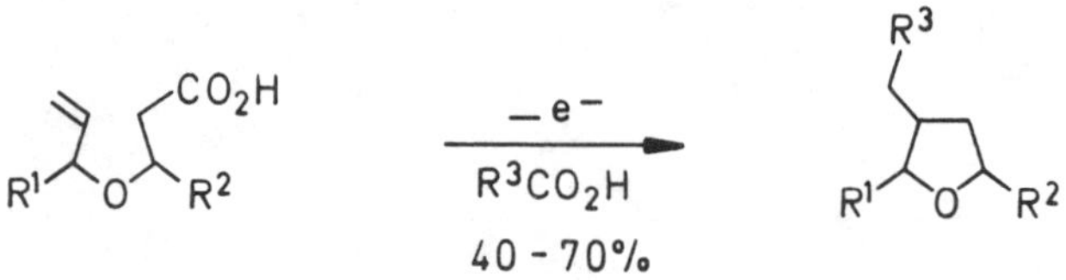

131

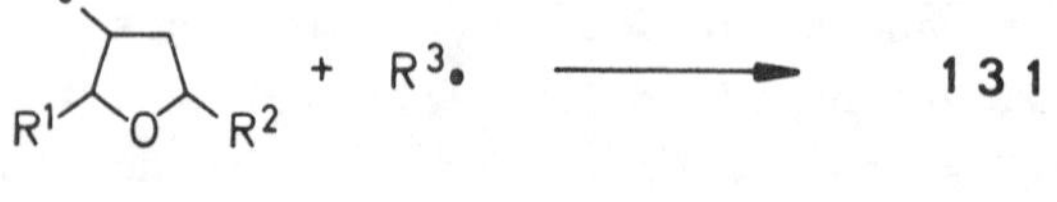

132

2. Reduction

Most of the precursors for radical cyclizations under reductive
conditions are ketones which form ketyl radical anions **133**.
Intramolecular additions to C-C multiple bonds give cyclized
radical anions **134** that are further reduced and protonated.
These reactions are not chain reactions and therefore require
an excess of reducing agent.

133   **134**

Pradhan[84] used alkali metals or sodium naphthalide for the
cyclization of 4,5-secocholestan-5-one **135**.

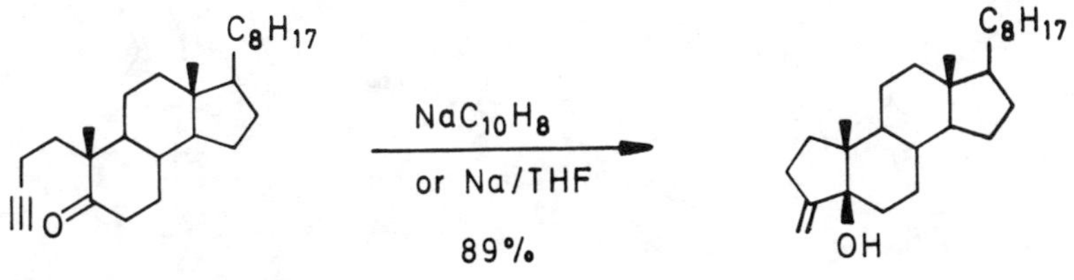

135

As a crucial step in the synthesis of gibberelic acid, Stork[85]
treated ketone **136** with Na in liquid ammonia and obtained the
cyclized product **137**.

136   **137**

Using Zn as the electron donor, Corey[86] trapped acyclic radicals with various functional groups.

In the formation of loganin tetraacetate **139** from secologanin tetraacetate **138** Hutchinson[87] used Mg as electron donor.

138   **139**

But it is not clear   whether, under cyclization conditions with Zn or Mg, the cyclized radicals abstract a hydrogen atom from the solvent or an electron from the metals. The acyloin condensation of the diester **140** gives **142**, presumably via ketyl radical **141**.[88]

140   **141**   **142**

Reduction of ketones and radical cyclizations can also be carried out in an electrochemical cell at the cathode.[89] Alkenes,[90] alkynes,[91] and allenes [92] have been used as C-C multiple bonds to trap ketyl radicals under these conditions.

ROS-G*

Photochemically induced electron transfer from tert. amines or hexamethylphosphortriamide to ketones can also lead to cyclized products.[93]

Pradhan[94] has demonstrated that not only ketones, but also
iodides can be reductively cyclized. Starting from **143**, the
iodide is generated in situ and cyclizes with Zn to give
norcholestane **144**.

143

X = OSO$_2$Me

144

E. Fragmentation and Displacement

Reactions in which radicals undergo fragmentation after
cyclization reactions have not been very frequently used in
synthesis. In one example, Keck[95] has applied the allyltin
method (see p. 98) to the synthesis of isoretronecanol **148**.
The radical precursor **146** is synthesized from allylthioether
145 via Bu$_3$SnH addition, followed by phenylthiyl elimination,
Mitsunobu coupling, reduction, and thiylation. After the
radical cyclization (**146→147**), oxidation of the olefinic bond,
and reduction of the amide yield the target molecule **148**.

145

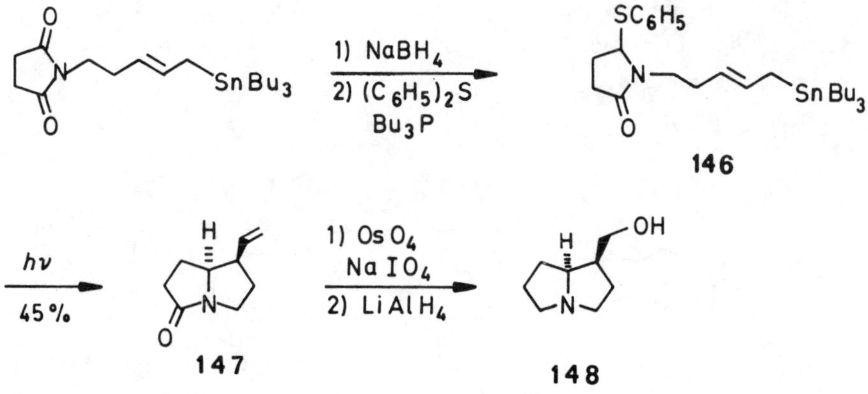

146

147   **148**

Carbon-halogen bonds ß to a radical center are also cleaved
rapidly; thus, the addition of CCl$_4$ to heptyne gives **152** as a
sideproduct   via radicals **149-151**.

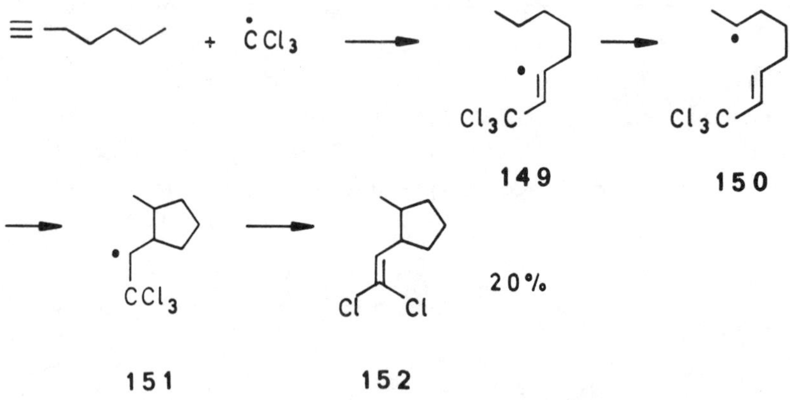

149   **150**

151   **152**   20%

The displacement of radicals can occur simultaneously with the formation of the ring if the bonds broken are very weak. Thus, Johnson[78,97] has shown that compounds **153** and **155** cleave the C-Br and C-Co bonds concomitant with the radical attack and yield cyclopropanes **154** and **156**, respectively.

153   **154**

155   **156**

With low concentrations of the intermolecular radical traps, cyclopentanes can also be formed. The thermal and photo-chemical reaction between 5-hexenylcobaloxime and a large excess of CCl_4 gives mainly the acyclic product **158**; whereas in the presence of a low concentration of CCl_4, cyclopentane **157** is formed.[78]

157   **158**

Replacement of a tin radical occurs during the formation of cyclopropane **160** from hydrazone **159**.[98] Radical **162** is the intermediate in this synthesis.

159   **160**   **161**

162

F. Biradicals

1. Reductive cyclization

Acyloin condensation is a synthetically important reductive cyclization reaction involving biradicals. The reaction is especially useful for the synthesis of medium-sized rings that are often difficult to make by other methods.[99]

Carbocycles, suffering from angle strain, can also be synthesized.[100]

2. Photolytical cyclization of ketones

Many photolytical C-C bond forming cyclizations such as the
di-π-methane rearrangement (**163→164**), the dienone-cyclo-
propylketone transformation (**165→166**), many intra- and intermolecular cycloadditions, as well as the Norrish I and
Norrish II reactions of carbonyl compounds proceed via biradicals.

It is far beyond the scope of this text to include these photo-
chemical reactions. There are several monographs and reviews
which give excellent overviews of photochemistry.[101] In this
chapter, mainly cyclizations are described in which biradicals
are generated by intramolecular H-abstraction reactions.

$$-\underset{|}{\overset{|}{C}}-(CH_2)_n-\overset{O}{\overset{\|}{C}}- \quad \xrightarrow{h\nu} \quad -\overset{\bullet}{\underset{|}{C}}-(CH_2)_n-\overset{\bullet}{\underset{|}{C}}-OH \quad \longrightarrow \quad -\underset{|}{\overset{|}{C}}-\underset{|}{\overset{OH}{\overset{|}{C}}}-$$

Cyclobutanes are formed following 1,5-hydrogen abstractions;
thus, methylisopulegon 167 gives the bicyclic compound 168,[102]
and the strained tricyclic compound 170 is synthesized from
169.[103]

167   168

169

170

1,6-Hydrogen abstraction leads to five-membered rings. Descotes[104] used this method for C-C bond formation reactions at the anomeric center of carbohydrates.

72 %

The pyrrolizidine alkaloid isoretronecanol 148 was synthesized from 171, which cyclizes to 172.[105]

171   172   148

Reactions involving 1,7-hydrogen abstraction form six-membered rings.

75%   25%

In the synthesis of alkaloids of the erythrina family, Mariano[107] cyclized the dihydroisoquinolinium salt **173** to **174**.

173   **174**

A biradical is presumably generated by a photostimulated electron transfer and desilylation step.

3. Diazenes

Nitrogen elimination of diazenes gives diradicals which
competitively undergo intramolecular radical coupling to yield
products of cyclization or disproportionation. Thus, from azo
compound **175** cyclobutanes **177** and alkene **178** are formed via
biradical **176.**[101]

	177a	**177b**	**178**
direct:	35%	3,5%	60%
sensitized:	11%	8%	77%

The amount of cyclized products **177** is relatively low and
depends on the condition of the photolysis. The formation of
three-membered rings **180** and **182** from pyrazoline derivatives
179 and **181** occurs with higher yields, although the product
ratio **182a:182b** remains dependent on the photolytic conditions.[108]

179   **180**   96 %

181   **182 a**   **182 b**

direct :   77 : 23
sensitized: 25 : 75

Sensitized photolysis leads to relatively long lived triplet
biradicals that loose the stereo- and regiochemical information
of the radical precursors **175** and **181**.

Little[109] elegantly applied trimethylenemethane biradicals
184,[110] generated from cyclic azo compound **183**, in the
synthesis of bi- and tricyclic target molecules.

183   **184**

Intramolecular diyl trapping of trimethylenemethane derivatives
leads to tricyclopentanoids of the hirsutene,[109,111] capnellene,[112] and coriolin[113] type. For the hirsutene synthesis,
thermolysis of diazene **185** in boiling acetonitrile gives
tricyclus **186**, which can be converted into hirsutene **55**.[111]

185   **186**   **55**

The formation of **186** from diazene **185** proceeds via biradical
187 which reacts with the neighboring double bond. The high
stereoselectivity of this reaction points to a preferred conformation, which could be caused by secondary orbital interactions between the ester group and the diyl ring carbon
atoms.

185 ⟶

187   (E = CO$_2$Me)   ⟶ **186**

The synthesis of coriolin **190** requires diazene **188**, which
upon photolysis, gives **189**.[112]

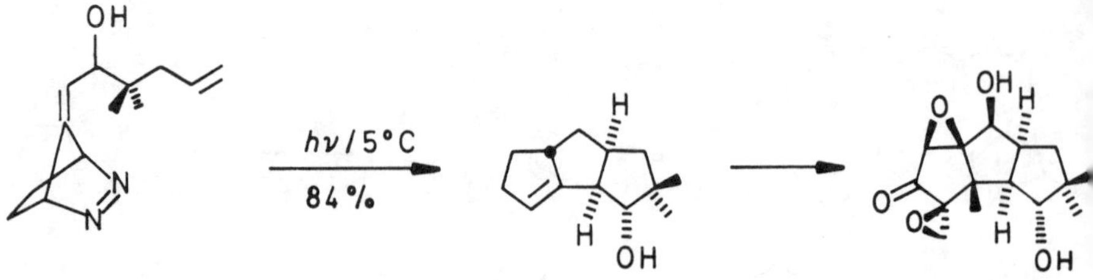

188   **189**   **190**

Detailed studies show that the stereochemistry of this diyl trapping reaction is temperature dependent.[114] Thus, the decomposition of diazene **191** gives products whose ratio changes by varying the reaction temperature.

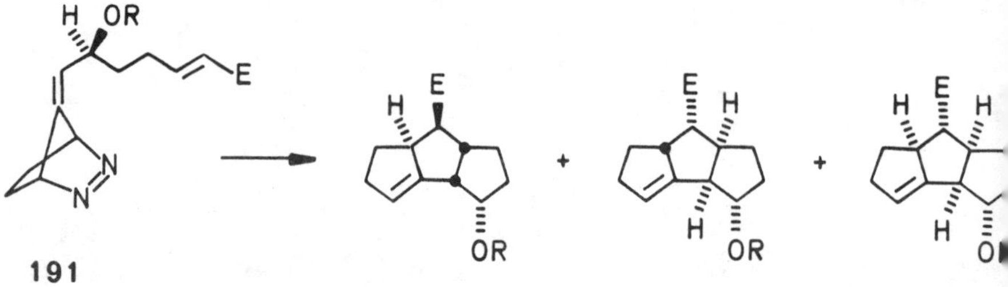

191

R : SiMe$_2$Bu^t

E : CO$_2$Me

-31°C	1	:	48,5	:	2,4
+81°C	1	:	11,7	:	0,8

Bicyclic cyclopentanoids can be synthesized if the multiple bond is conjugated with the radical center. In these cases, even C-O double bonds can be used in direct combination reactions of diradical **192**.[115]

192

The trimethylenemethane derivatives **193** can also be trapped intermolecularly by alkenes, alkynes, aldehydes, ketones, imines, and thioketones.[116]

The low regioselectivity of these reactions limits their synthetic usefulness. Therefore, although the precursors are easily accessible and the total yields are high, the synthesis of complex tricyclopentanoids via intermolecular diyl trapping cannot compete with the intramolecular method.

REFERENCES

1. D.J. Hart, *Science* **1984**, *223*, 883; B. Giese, *Angew. Chem.*
    Int. Ed. Engl. **1985**, *24*, 553, B. Giese (ed.): *Selectivity*
    and Synthetic Applications of Radical Reactions, Tetra-
    hedron **1985**, *41*, 3887.

2. P. Schmid, D. Griller, K.U. Ingold, *Int. J. Chem. Kinet.*
    1979, *11*, 333; P.I. Abell in C.H. Bamford, C.F.H. Tipper
    (ed.): *Kinetics* Vol. XVIII, Elsevier, Amsterdam 1976,
    p. 111.

3. C. Chatgilialoglu, K.U. Ingold, J.C. Scaiano, *J. Am. Chem.*
    Soc. **1981**, *103*, 7739.

4. S.U. Park, S.K. Chung, M. Newcomb, *J. Am. Chem. Soc.* **1986**,
    108, 240.

5. A.L.J. Beckwith, K.U. Ingold in P. de Mayo (ed.):
    Rearrangements in Ground and Excetid States Vol. 1,
    Academic Press, New York 1980, p. 161; J.M. Surzur in
    R.A. Abramovitch (ed.): *Reactive Intermediates*, Vol. 2,
    Plenum Press, New York 1982, p. 121.

6. D.I. Davies in D.H. Hey, W.A. Waters, R.O.C. Norman (ed.):
    Essays on Free-Radical Chemistry Burlington House, London
    1970, p. 201; B. Giese, K. Jay, *Chem. Ber.* **1977**, *110*,
    1364; P.C. Wong, D. Griller, *J. Org. Chem.* **1981**, *46*, 2327.

7. T.A. Halgren, M.E.H. Howden, M.E. Medorf, J.D. Roberts,
    J. Am. Chem. Soc. **1967**, *89*, 3051.

8. G. Greig, J.C.J. Thynne, *Trans Faraday Soc.* **1966**, *62*,
    3338; S. Sustmann, C. Rüchardt, *Chem. Ber.* **1975**, *108*, 3043.

9. S.J. Christol, R.J. Daughenbaugh, *J. Org. Chem.* **1979**, *44*,
    3434.

10. A.L.J. Beckwith, C.H. Schiesser, *Tetrahedron* **1985**,*41*, 3925;
    A.L.J. Beckwith, G.F. Meijs, *J. Chem. Soc. Perkin Trans. 2*
    1979, 1535; A.L.J. Beckwith, G. Phillipou, A.K. Serelis,
    Tetrahedron Lett. **1981**, *22*, 2811.

11. N.A.Porter, unpublished results.

12. J.E. Baldwin, *J. Chem. Soc. Chem. Commun.* **1976**, 734.

13. B. Giese, *Angew. Chem. Int. Ed. Engl.* **1983**, *22*, 753.

14. M. Julia, C. Descoins, M. Baillarge, B. Jacquet, D. Uguen, F.A. Groeger, *Tetrahedron* **1975**, *31*, 1737.

15. P. Bischof, *Tetrahedron Lett.* **1979**, 1291.

16. A.N. Abeywickreyma, V.W. Bowry, S.A. Glover, P.E. Pigou, unpublished results.

17. M. Julia, *Acc. Chem. Res.* **1971**, 4, 386; M. Julia, *Pure Appl. Chem.* **1974**, *40*, 553.

18. J.W. Wilt, *Tetrahedron* **1985**, *41*, 3979.

19. M.D. Bachi, F. Frolow, C. Hoornaert, *J. Org. Chem.* **1983**, *48*, 1841.

20. D.A. Burnett, J.K. Choi, D.J. Hart, Y.M. Tsai, *J. Am. Chem. Soc.* **1984**, *106*, 8201.

21. A.L.J. Beckwith, C.J. Easton, A.K. Serelis, *J. Chem. Soc. Chem. Commun.* **1980**, 482.

22. E.J. Corey, C. Shih, N.Y. Shih, K. Shimoji, *Tetrahedron Lett.* **1984**, *25*, 5013.

23. D.G. Wujek, N.A. Porter, *Tetrahedron* **1985**, *41*, 3973.

24. S. Danishefsky, S. Chackalamannil, B.J. Uang, *J. Org. Chem.* **1982**, *47*, 2231; S. Danishefsky, E. Taniyama. *Tetrahedron Lett.* **1983**, *24*, 15.

25. Y. Matsuki, M. Kodama, S. Ito, *Tetrahedron Lett.* **1979**, 4081.

26. P. Bakuzis, O.S. Campos, M.L.F. Bakuzis, *J. Org. Chem.* **1976**, *41*, 3261; G. Büchi, H. Wüest, *J. Org. Chem.* **1979**, *44*, 546.

27. G. Stork in H. Nozaki (ed.): *Current Trends in Organic Synthesis*, Pergamon Press, Oxford 1983, p. 359.

28. G. Stork, R. Mook, *J. Am. Chem. Soc.* **1983**, *105*, 3721.

29. D.P. Curran, D.M. Rakiewicz, *Tetrahedron* **1985**, *41*, 3943.

30. D.P. Curran, M.H. Chen, *Tetrahedron Lett.* **1985**, *26*, 4991.

31. D.P. Curran, S.C. Kuo, *J. Am. Chem. Soc.* **1986**, *108*, 1106.

32. A.L.J. Beckwith, D.H. Roberts, C.H. Schiesser, A. Wallner, *Tetrahedron Lett.* **1985**, *26*, 3349.

33. W.R. Leonard, T. Livinghouse, *Tetrahedron Lett.* **1985**, *26*, 6431.

34. C.P. Chuang, D.J. Hart, *J. Org. Chem.* **1983**, *48*, 1782; D.J. Hart, H.C. Huang, *Tetrahedron Lett.* **1985**, *26*, 3749.

35. J.K. Choi, D.J. Hart, *Tetrahedron* **1985**, *41*, 3959.

36. M.D. Bachi, C. Hoornaert, *Tetrahedron Lett.* **1982**, *23* 2505.

37. M.D. Bachi, C. Hoornaert, *Tetrahedron Lett.* **1981**, *22*, 2689.

38. J.J. Köhler, W.N. Speckamp, *Tetrahedron Lett.* **1977**, 635.

39. A. Padwa, H. Nimmesgern, G.S.K. Wong, *J. Org. Chem.* **1985**, *50*, 5620.

40. C.S. Wilcox, L.M.Thomasco, *J. Org. Chem.* **1985**, *50*, 546.

41. G. Stork, N.H. Baine, *Tetrahedron Lett.* **1985**, *26*, 5927.

42. N.N. Marinovic, H.Ramanathan,*Tetrahedron Lett.* **1983**, *24*, 1871.

43. G. Stork, R. Mook, S.A. Biller, S.D. Rychnovsky, *J. Am. Chem. Soc.* **1983**, *105*, 3741.

44. D.L.J.Clive,P.L. Beaulieu, *J. Chem. Soc. Chem. Commun.* **1983**, 307.

45. M.D. Bachi, E. Bosch, *Tetrahedron Lett.* **1986**,*27*, 641.

46. M. Ladlow, G. Pattenden, *Tetrahedron Lett.* **1984**, *25*, 4317.

47. F. Flies, R. Lalande, B. Maillard, *Tetrahedron Lett.* **1976**, 439.

48. R. Gompper, D. Lach, *Tetrahedron Lett.* **1973**, 2687; M. Apparu, J.K. Crandall, *J. Org. Chem.* **1984**, *49*, 2125.

49. D.D. Tanner, P.M. Rahimi, *J. Org. Chem.* **1979**, *44*, 1674.

50. D.L.J. Clive, P.L. Beaulieu, L. Set, *J. Org. Chem.* **1984**, *49*, 1313.

51. G.A. Kraus, Y.S. Hon, *J. Org. Chem.* **1985**, *50*, 4605.

52. N. Ono, H. Miyake, A. Kaminura, I. Hamamoto, R. Tamura,
     A. Kaji, *Tetrahedron* **1985**, *41*, 4013.

53. G. Stork, P.M. Sher, *J. Am. Chem. Soc.* **1986**, *108*, 302.

54. M. Julia, J.M. Surzur, L. Katz, *C.R. Acad. Sci. Ser. C*
     1960, *251*, 1030; M. Julia, *Rec. Chem. Progr.* **1964**, *25*, 3;
     M. Julia, *Pure Appl. Chem.* **1967**, *15*, 167.

55. M. Julia, C. James, *C. R. Acad. Sci. Ser. C* **1962**, *255*, 959.

56. M. Julia, F. Le Goffic *Bull. Soc. Chim. Fr.* **1964**, 1129;
     J.D. Winkler, V. Sridar, *J. Am. Chem. Soc.* **1986**, *108*, 1708.

57. J.P. Montheard, *C. R. Acad. Sci. Ser. C* **1965**, *260*, 577.

58. D.H.R. Barton, D.L.J. Clive, P.D. Magnus, G. Smith,
     J. Chem. Soc. C **1971**, 2193.

59. E. van Bruggen, *Rec. Trav. Chim. Pays-Bas* **1968**, *87*, 1134.

60. D. Elad in O.L. Chapman: *Organic Photochemistry*, Vol. 2,
     Marcel Dekkar, New York 1969, p. 168.

61. L. Freedman, *J. Am. Chem. Soc.* **1964**, *86*, 1885; P. Gott-
     schalk, D.C. Neckers, *J. Org. Chem.* **1985**, *50*, 3498.

62. T.W. Sam, J.K. Sutherland, *J. Chem. Soc. Chem. Commun.*
     1971, 970; J.G. Traynham, H.H. Hsieh, *J. Org. Chem.* **1973**,
     38, 868; E.D. Brown, T.W. Sam, J.K. Sutherland, A. Torre
     J. Chem. Soc. Perkin Trans. 1 **1975**, 2326.

63. N.O. Brace, *J. Org. Chem.* **1966**, *31*, 2879; N.O. Brace,
     J. Org. Chem. **1973**, *38*, 3167.

64. P. Piccardi, M. Modena, L. Cavalli, *J. Chem. Soc. C* **1971**,
     3959; P. Piccardi, P. Massardo, M. Modena, E. Santoro,
     J. Chem. Soc. Perkin Trans. 1 **1974**, 1848.

65. J.G. Traynham, H.H. Hsieh, *J. Org. Chem.* **1973**, *38*, 868.

66. S. Takano, S. Nishizawa, M. Akiyama, K. Ogasawara,
     Synthesis **1984**, 949.

67. H. Nagashima, K. Ara, H. Wakamatsu, K. Itoh, *J. Chem.
     Soc. Chem. Commun.* **1985**, 518.

68. D.P. Curran, M.H. Chen, D. Kim, *J. Am. Chem. Soc.* **1986**,
     108, 2489.

69. G.A. Kraus, K. Landgrebe, *Tetrahedron* **1985**, *41*, 4039.

70. D.H.R. Barton, D. Crich, G. Kretzschmar, *J. Chem. Soc.
     Perkin Trans. 1* **1986**, 39.

71. M.E. Kuehne, R.E. Damon, *J. Org. Chem.* **1977**, *42*, 1825.

72. N.A. Porter, M.O. Funk, *J. Org. Chem.* **1975**, *40*, 3614.

73. M. Okabe, M. Tada, *Chemistry Lett.* **1980**, 831.

74. M. Ladlow, G. Pattenden, *Tetrahedron Lett.* **1984**, *25*, 4317.

75. S. Torii, T. Inokuchi, T. Yukawa, *J. Org. Chem.* **1985**, *50*,
     5875.

76. M. Okabe, M. Abe, M. Tada, *J. Org. Chem.* **1982**, *47*, 1775,
     5382.

77. R. Scheffold, M. Dike, S. Dike, T. Herold, L. Walder,
     J. Am. Chem. Soc. **1980**, *102*, 3642; R. Scheffold, *Chimica*
     1985, *39*, 203.

78. M.D. Johnson, *Acc. Chem. Res.* **1983**, *16*, 343.

79. E.J. Corey, M. Kang, *J. Am. Chem. Soc.* **1984**, *106*, 5384.

80. A.B. Ernst, W.E. Fristad, *Tetrahedron Lett.* **1985**, *26*, 3761.

81. B.B. Snider, R. Mohan, S.A. Kates *J. Org. Chem.* **1985**, *50*,
     3661.

82. R. Breslow, J.T. Groves, S.S. Olin *Tetrahedron Lett.* **1966**,
     4717; R.M. Coates, L.S. Melvin, *J. Org. Chem.* **1970**, *35*,
     865.

83. R. Breslow, S.S. Olin, J.T. Groves, *Tetrahedron Lett.*
     1968, 1837.

84. S.K. Pradhan, S.R. Kadam, J.N. Kolhe, T.V. Radhakrishnan,
     S.V. Sohani, V.B. Thaker, *J. Org. Chem.* **1981**, *46*, 2622.

85. G. Stork, R.K. Boeckmann, D.F. Taber, W.C. Still, J. Singh,
     J. Am. Chem. Soc. **1979**, *101*, 7107.

86. E.J. Corey, S.G. Pyne, *Tetrahedron Lett.* **1983**, *24*, 2821.

87. T. Ikeda, S. Yue, C.R. Hutchinson, *J. Org. Chem.* **1985**,
     50, 5193.

88.   P.Y. Johnson, M.A. Priest, *J. Am. Chem. Soc.* **1974**, *96*, 5618.

89.   H.J. Schäfer, *Angew. Chem. Int. Ed. Engl.* **1981**, *20*, 911.

90.   T. Shono, I. Nishiguchi, H. Ohmizu, M. Mitani, *J. Am. Chem. Soc.* **1978**, *100*, 545.

91.   R. Pallaud, M. Nicolaus, *C. R. Acad. Sci. Ser.* **1968**, *C 267*, 1834; M.J. Allen, J.A. Siragusa, W. Pierson, *J. Chem. Soc.* **1960**, 1045.

92.   G. Pattenden, G.M. Robertson, *Tetrahedron* **1985**, *41*, 4001.

93.   D. Belotti, J. Cossy, J.P. Pete, C. Portella, *Tetrahedron Lett.* **1985**, *26*, 4591.

94.   S.K. Pradhan, J.N. Kolhe, J.S. Mistry, *Tetrahedron Lett.* **1982**, *23*, 4481.

95.   G.E. Keck, E.J. Enholm, *Tetrahedron Lett.* **1985**, *26*, 3311.

96.   E.I. Heiba, R.M. Dessau, *J. Am. Chem. Soc.* **1967**, *89*, 3772; K.M. Kopchik, J.A. Kampmeier, *J. Am. Chem. Soc.* **1968**, *90*, 6733.

97.   A. Bury, S.T. Corker, M.D. Johnson, *J. Chem. Soc. Perkin Trans. 1* **1982**, 645.

98.   H. Nishiyama, H. Arai, Y. Kanai, H. Kawashima, K. Itoh, *Tetrahedron Lett.* **1986**, *27*, 361.

99.   J.J. Bloomfield, D.C. Owsley, J.M. Nelke, *Org. Reactions* **1976**, *23*, 259.

100. J.J. Bloomfield, *Tetrahedron Lett.* **1968**, 587; J.M. Conia, J.M. Denis, *Tetrahedron Lett.* **1969**, 3545.

101. D.O. Cowan, R.L. Drisko: *Elements of Organic Photochemistry*, Plenum Press, New York 1976.

102. R.C. Cookson, J. Hudec, A. Szabo, G.E. Usher, *Tetrahedron* **1968**, *24*, 4353.

103. A. Padwa, W. Eisenberg, *J. Am. Chem. Soc.* **1970**, *92*, 2590.

104. C. Bernasconi, L. Cottier, G. Descotes, J.P. Praly, G. Remy, M.F. Grenier-Loustalot, F. Metras, *Carbohydr. Res.* **1983**, *115*, 105.

105. J.C. Gramain, R. Remuson, D. Vallée, J. Org. Chem. **1985**, 50, 710.

106. M. Barnard, N.C. Yang, Proc. Chem. Soc., London **1958**, 302.

107. R.Ahmed-Schofield, P.S. Mariano, J. Org. Chem. **1985**, 50, 5667.

108. S.D. Andrews, A.C. Day, J. Chem. Soc. Chem. Commun. **1966**, 667; T. Aratani, Y. Nakanisi, H. Nozaki, Tetrahedron **1970**, 26, 4339.

109. R.D. Little, G.W. Muller, J. Am. Chem. Soc. **1979**, 101, 7129.

110. J.A. Berson, Acc. Chem. Res. **1978**, 11, 446.

111. R.D. Little, G.W. Muller, J. Am. Chem. Soc. **1981**, 103, 2744.

112. R.D. Little, G.L. Carroll, Tetrahedron Lett.**1981**, 22, 4389.

113. L. Van Hijfte, R.D. Little, J. Org. Chem. **1985**, 50, 3940.

114. K.J. Stone, R.D. Little, J. Am. Chem. Soc. **1985**, 107, 2495.

115. K.D. Moeller, R.D. Little, Tetrahedron Lett. **1985**, 26, 3417.

116. R.D. Little, H. Bode, K.J. Stone, O. Wallquist, R. Dannecker,J. Org. Chem. **1985**, 50, 2400.

117. R.D. Little, A. Bukhari, M.G. Venegas, Tetrahedron Lett. **1979**, 305.

Chapter 5

C—C Bond Formation of Aromatic Systems

A. Introduction

Formation of C-C bonds by radical reactions involving neutral or charged, carbocyclic or heterocyclic aromatic systems can occur via radicals **1 - 4**.

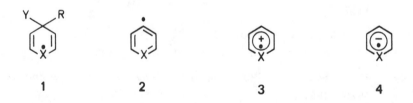

1　　　**2**　　　**3**　　　**4**

Radical **1** corresponds to the Wheland or Meisenheimer complex of ionic substitution reactions and is formed by radical attack at aromatic compounds. The aromaticity is regained by splitting off Y·.

1

Aryl radicals **2**, which have not lost aromaticity during their formation, are very reactive σ-radicals that rapidly attack unsaturated neutral or charged carbon atoms. They can be generated by several routes, including electron transfer reactions. The reaction products are formed from radical **5** via hydrogen abstraction, heteroatom donation, electron transfer, or fragmentation.

2   **5**

Radical ions **3** are formed by one-electron transfer reactions from the neutral aromatic compounds and react either with ions (**3** → **6**) or with radicals (**3** → **7**). The adduct radicals **6** and ions **7** yield products via hydrogen and proton abstraction, respectively.

B. Reactions with Carbocyclic Aromatic Compounds

Benzene behaves like an electron-rich alkene and is attacked
by nucleophilic alkyl radicals with rate coefficients of
$10 - 10^3$ (1/mol·s) at 25-80°C.[1,2] This is slightly slower
than the addition to alkylated alkenes[2,3] and is hardly
fast enough for synthetic applications (see p. 12). However,
electrophilic radicals such as **9**, or σ-radicals **10** and **11**
are reactive enough[4] to be used in syntheses.

Nucleophilic π-radicals **8** are successful only if the aromatic compounds are substituted with electron-withdrawing groups or if electron-poor heterocyclic salts are used.[5]

1. Electrophilic radicals

Electron-rich aromatic compounds are readily attacked by electrophilic nitrogen- or oxygen-centered radicals generated from protonated N-chloroamines[5,6] and peroxides,[7] respectively. These syntheses involve radical chain reactions in which cyclohexadienyl radicals are trapped either by halogen abstraction or electron transfer.

Carbon centered radicals exhibit electrophilic behavior if they are substituted by electron-withdrawing groups; this is clearly demonstrated by the ρ^+-values of Table 1.[8]

<u>Table 1:</u> Comparison of ρ^+-values of homolytic
aromatic substitution by radicals X·

X·	ρ^+
C_6H_{11}·	1.1
CH_3·	0.1
C_6H_5·	0.1
H_3CCOCH_2·	-1.5
O_2NCH_2·	-2.1
$C_6H_5CO_2$·	-1.6
$i-C_3H_7OCO_2$·	-2.3

Because of their negative ρ^+-values, acyl- and nitro-substituted radicals are electrophiles and have found synthetic
applications in C-C bond forming reactions with aromatic
compounds. The radicals are generated by oxidation of
acetone[8] or nitromethane[9] with Mn^{3+} or Ce^{4+} salts (see p. 89).

$$CH_3-\overset{\overset{O}{\|}}{C}-CH_3 \;\rightleftharpoons\; CH_3-\overset{\overset{OH}{|}}{C}=CH_2 \;\xrightarrow{Mn^{3+}}\; CH_3-\overset{\overset{OH}{|}}{C}\overset{+\bullet}{-} CH_2$$

$$\xrightarrow{-H^+} CH_3-\overset{\overset{O}{\|}}{C}-\overset{\bullet}{C}H_2$$

$$CH_3-NO_2 \quad \rightleftharpoons \quad CH_2 = \overset{+}{N} \underset{OH}{\overset{O^-}{\diagdown}} \quad \xrightarrow{Mn^{3+}} \quad \overset{\bullet}{C}H_2 - \overset{+}{N} \underset{OH}{\overset{O}{\diagdown}}$$

$$\xrightarrow{-H^+} \quad \overset{\bullet}{C}H_2-NO_2$$

The yields are higher with anisole or toluene than with halobenzenes because of the electrophilic nature of acyl- and nitro-substituted radicals.

X : OCH$_3$ (74%; o:m:p = 84:3:13); CH$_3$ (51%; o:m:p = 66:20:14)

H (40%); F (29%; o:m:p = 71:9:20)

X : OCH$_3$ (77%; o:m:p = 71:5:24); CH$_3$ (77%; o:m:p = 52:27:21)

H (78%); Cl (20%; o:m+p = 52:48)

These radical alkylation reactions are not chain reactions because the formation of the attacking radicals, as well as the reaction of the cyclohexadienyl radicals, require stoichiometric amounts of Mn^{3+} or Ce^{4+} salts as oxidants.

$$CH_3Y \xrightarrow{Mn^{3+}} \overset{\bullet}{C}H_2Y$$

2. σ-Radicals

Besides electrophilic radicals, σ-radicals are also reactive enough to add to benzoic systems with synthetically useful rates. The phenyl radical attacks benzene with a rate coefficient k = $4.5 \cdot 10^5$ (1/mol·s) at $25°C$;[4] therefore, arylation of aromatic compounds can be carried out successfully. Actually, the phenylation of benzene using N-nitrosoacetanilide 12 was discovered by Bamberger[10] in 1896. Forty years later, Hey[11] proposed that the phenyl radical is a reaction intermediate.

$$C_6H_5\overset{NO}{\underset{O}{N-\overset{\parallel}{C}CH_3}} + C_6H_6 \longrightarrow C_6H_5-C_6H_5 + N_2 + CH_3CO_2H$$

12

This reaction played a major part in the development of the
concept and understanding of the transient existence of
radicals in solution. From the extensive work of Hey,[11,12]
Huisgen,[13] Rüchardt,[14] Perkins,[15] and Cadogan,[16] it seems that
the following mechanism might be operative: The key step is the
attack of the phenyl radical on benzene to give adduct radical
13, which is oxidized either by nitroxide radicals or diazonium
ions to the cyclohexadienyl cation **14**; deprotonation then
gives biphenyl.

13   **14**

N-Nitrosoacetanilides **15**,[17] diazonium salts **16**,[18] azo compounds
17,[19] triazenes **18**,[20] diacyl peroxides **19**,[21] carboxylic acids
20,[22] arenecarbonyloximes **21**,[23] and arylhalides **22**[24] can be
used as precursors for aryl radicals.

15   **16**   **17**   **18**

19   **20**   **21**   **22**

Radical arylations of aromatic compounds are more of mechanistic than of synthetic interest because in most cases
positional and substrate selectivities (regioselectivity and
chemoselectivity) are quite low. Generally, all of the free
positions of an aromatic substrate are substituted,[25] and at
higher conversion polysubstitution occurs.

$C_6H_5\bullet$ + [benzene ring with OCH_3] $\longrightarrow$ H_5C_6—[benzene ring with OCH_3]

o : m : p = 69 : 18 : 13

$C_6H_5\bullet$ + [benzene ring with NO_2] $\longrightarrow$ H_5C_6—[benzene ring with NO_2]

o : m : p = 62 : 8 : 30

The ρ-values of table 1 (see p. 214) show that this low
selectivity is due to the small polar effects of arylation
reactions.

Intramolecular reactions are governed less by polar than by
ring strain effects. Therefore, intramolecular arylation,
like the Pschorr synthesis of phenanthrene systems, occurs
with higher selectivity. The cyclization **23** → **24**, which is the
key step in the synthesis of the alkaloid tylocrebrine **25**,[26]
and also the formation of spiro compound **27** by photolysis
of bromide **26**[27] demonstrate this selectivity.

Tiecco[28] has shown that in reactions with σ-radicals such as the 1-adamantyl radical **11** ipso substitution can also occur. Thus, nitro, acyl, nitrile, and sulfonyl groups are substituted by σ-radicals if the aromatic compounds are made strongly electron-deficient by the presence of other electron-with-drawing substituents.

$X = NO_2, CN, COR, CHO, CO_2R, SO_2R$

3. Nucleophilic π-radicals

Substitution reactions of nucleophilic alkyl radicals with π-character are synthetically important mainly in reactions with protonated heterocyclic bases. Synthetic applications with benzoid compounds are rare because addition rates and position selectivities are low.

Therefore, only a few synthetically interesting cases exist in which alkyl radicals of this type, generated from for example iodides, react with benzoid systems.[29,30]

$CH_2=CH-CH_2I$ +   [OCH_3 phenol] $\xrightarrow[44\%]{h\nu}$ [allyl-OCH_3 product]

28   **29**

o : m : p = 48 : 16 : 36

30   **31a**   **31b**

It is believed that the synthesis of **29** and **31** proceeds via attack of the alkyl radical at the aromatic ring.[30] The formation of **31b** is an example of an intramolecular ipso substitution.

C. Reactions with Heterocyclic Aromatic Compounds

Alkylations and acylations of heterocyclic aromatic compounds developed by Minisci[5,6] and his group are very useful substitution reactions that proceed by radical addition at the aromatic ring. These reactions are equivalent to Friedel-Crafts alkylations and acylations, but in contrast to Friedel-Crafts reactions, they are carried out with electron-poor aromatic systems. Thus, radical chemistry makes possible C-C bond formation reactions which are difficult to accomplish using ionic methods. The rates are particularly high in the presence of strong acids which protonate heterocyclic bases. Prim. butyl radicals attack protonated pyridine with a rate coefficient of $4.4 \cdot 10^4$ (1/mol·s) at 57°C.[31] Electron-withdrawing groups, annulation with aromatic rings, and introduction of heteroatoms in the ring increase the reactivity. In an analogous manner, electron-donating substituents at the radical center, e.g. alkyl groups, in most cases raise the reactivity of radicals.[31]

$k_{C_3H_7CH_2 \bullet}$ $4.4 \cdot 10^4$ $8.9 \cdot 10^5$ $8.5 \cdot 10^5$ $1.8 \cdot 10^7$ (l/mol·s)

$k_{(CH_3)_3C \bullet}$ $3.3 \cdot 10^4$ $6.3 \cdot 10^7$ $4.1 \cdot 10^6$ (l/mol·s)

These substituent effects demonstrate the nucleophilic behavior of alkyl radicals in their reactions with protonated heterocyclic aromatic compounds. The polar effects can be explained by an interaction between the SOMO of the radical and the LUMO of the aromatic compounds (see p. 15).

Acyl radicals are σ-radicals which also behave like nucleophiles and attack protonated pyridines with a rate coefficient of 10^5 - 10^6 (1/mol·s).[32] Therefore, not only alkylation, but also acylation of heteroaromatic compounds can be successfully carried out.

The chain reaction proceeds via addition of radical **32** at the protonated heteroaromatic ring **33**, to give the azacyclohexadienyl radical **34** which is oxidized, e.g. by metal ions, to give product **35**. The reduced metal ion regenerates the starting radical **32** by reaction with suitable precursors.

The rearomatization of **34** proceeds presumably via radical **36**, which is rapidly oxidized to **35**.[33]

In general, the regioselectivity increases with increasing polar effects. Thus, tert-butyl radicals are slightly more selective than cyclohexyl and ethyl radicals.[6]

$$R = C_2H_5 \qquad 56 \quad : \quad 44$$
$$c - C_6H_{11} \qquad 63 \quad : \quad 37$$
$$t - C_4H_9 \qquad 68 \quad : \quad 32$$

Very high regioselectivities are achieved if the heteroaromatic compounds are further substituted or annulated with aromatic rings.[6]

1. Alkylation

A large variety of unsubstituted and substituted alkyl radicals are readily available from hydrocarbons, alkyl halides, alcohols, ethers, ketones, acids, and formamides.[6]

a. Hydrocarbons

Hydrogen abstraction from unactivated hydrocarbons can be achieved with dialkylammonium radical cations, which are generated from chloroammonium ions with Fe^{2+} salts.[34]

$$R-H \quad + \quad \underset{\overset{|}{H}}{\underset{N^+}{\bigcirc}}^{X} \quad + \quad R_2\overset{+}{N}HCl \quad \xrightarrow[Fe^{2+}]{H_2SO_4} \quad \underset{\overset{|}{H}}{\underset{N^+}{\bigcirc}}_R^{X}$$

$$R_2\overset{+}{N}HCl \quad + \quad Fe^{2+} \quad \longrightarrow \quad R_2\overset{+}{\overset{\bullet}{N}}H \quad + \quad Fe^{3+} \quad + \quad Cl^-$$

$$R_2\overset{+}{\overset{\bullet}{N}}H \quad + \quad R-H \quad \longrightarrow \quad R\bullet \quad + \quad R_2\overset{+}{N}H_2$$

It is believed that the cyclohexadienyl radical cation **34**, formed after radical attack, abstracts a chlorine atom from the chloroammonium ion and yields product **35** by dehydro-chlorination.[34]

$$\underset{\overset{|}{H}}{\underset{N^+}{\bigcirc}}_R^{X} \quad \xrightarrow{R_2\overset{+}{N}HCl} \quad \underset{\overset{|}{H}}{\underset{N^+}{\bigcirc}}_R^{X} \quad \longrightarrow \quad \underset{\overset{|}{H}}{\underset{N^+}{\bigcirc}}_R^{X}$$

34   **35**

The difficulty with hydrocarbons lies in the regioselectivity of the hydrogen abstraction step. With cyclic hydrocarbons like cyclohexane there is, of course, no selectivity problem, and in acyclic systems the selectivity question can be solved by introducing electron-withdrawing substituents, which direct the attack of the electrophilic radical to the ω-1 position.[34] Thus, the acyclic hydrocarbons **37** react with high regioselectivities that increase with the bulk of the substituents R at the nitrogen centered radicals.[34,35]

$$\text{(pyrazine)} + C_6H_{12} + (CH_3)_2NCl \xrightarrow[H_2SO_4]{Fe^{2+}} \text{(pyrazine)}-C_6H_{11} \quad 45\%$$

$$CH_3-CH_2(CH_2)_n Y + R_2\overset{\bullet}{\overset{+}{N}}H \longrightarrow CH_3-\overset{\bullet}{C}H(CH_2)_n Y$$

37

$$Y = CO_2R, \ Cl, \ OCH_3, \ NH_2, \quad n = 3,4$$

b. Alkyl iodides

In the presence of dibenzoyl peroxide, alkyl iodides are suitable precursors for radical substitution reactions of heteroaromatic compounds.[36]

$$\text{(4-cyanopyridine)} + \text{iPr}-I + (C_6H_5CO_2)_2 \xrightarrow[\substack{72\% \\ \text{conversion}}]{Fe^{3+}/H^+} \text{(2-iPr-4-CN-pyridine)} + \text{(2,6-diiPr-4-CN-pyridine)}$$

66%   34%

$$\text{(2-methylquinoline)} + C_6H_{11}I + (C_6H_5CO_2)_2 \xrightarrow[\substack{92\% \\ \text{conversion}}]{Fe^{3+}/H^+} \text{(4-cyclohexyl-2-methylquinoline)} \qquad 88\%$$

$$\text{(benzothiazole)} + \text{(i-Pr)}{-}I + (C_6H_5CO_2)_2 \xrightarrow[\substack{66\% \\ \text{conversion}}]{Fe^{3+}/H^+} \text{(2-isopropylbenzothiazole)} \qquad 90\%$$

Dibenzoyl peroxide oxidizes radical **34** to the product **35**. The benzoyloxy radical, thus formed, loses CO_2 and gives phenyl radicals which abstract iodine from alkyl iodides.[35]

$$\textbf{34} + (C_6H_5CO_2)_2 \longrightarrow \textbf{35} + C_6H_5CO_2{}^{\bullet} + C_6H_5CO_2{}^{-}$$

$$C_6H_5CO_2{}^{\bullet} \xrightarrow{-CO_2} C_6H_5{}^{\bullet} \xrightarrow{R-I} R^{\bullet} + C_6H_5I$$

c. Alcohols and ethers

α-Oxyalkyl radicals can be easily obtained by hydrogen abstraction from alcohols and ethers with redox systems like H_2O_2/M^{n+}, $t\text{-BuOOH}/M^{n+}$, $\overset{+}{N}H_3OSO_3^-/M^{n+}$, $S_2O_8{}^{2-}/M^{n+}$ ($M^{n+}=Fe^{2+}$, Ti^{3+}), $\overset{+}{N}H_3OH/Ti^{3+}$, or oxidants like $S_2O_8{}^{2-}$ and dibenzoyl peroxide.[37]

Intermediates which abstract hydrogen atoms from alcohols and ethers are heteroatom centered radicals or phenyl radicals.

The formation of oxygen centered radicals from peroxides is a well known reaction that occurs via electron transfer from the metal ion to the peroxide and cleavage of the O-O bond. The oldest reagent of this type is Fenton's reagent (H_2O_2/ Fe^{2+}).[38]

$$RO-OR \;+\; M^{n+} \longrightarrow RO^{\bullet} \;+\; RO^{-} \;+\; M^{(n+1)+}$$

These very reactive radicals attack not only α-CH bonds of alcohols and ethers, a considerable amount results also from abstraction of more remote hydrogen atoms. The regioselectivity depends very much on the reaction conditions.[37]

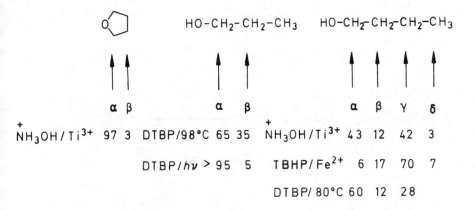

A problem also arises from the oxidizability of alkoxyalkyl radicals, especially if further alkyl groups make the oxidation even easier. Therefore, mild oxidants should be used in these cases, but their oxidizing power must be large enough for the rearomatization of the heterocyclic intermediate **36**. The $\overset{+}{N}H_3OH/Ti^{3+}$ redox system fulfills these requirements.[37]

d. Alkenes

Electron-rich alkenes can be used as radical precursors because they are easily attacked by electrophilic radicals Y• to give nucleophilic alkyl radicals **38**, which react with heterocycles in acidic solution to give substitution products **39**.

38   **39**

Electrophilic radicals are generated by oxidation of ketones or azides.[6,39]

45 %

Treatment of alkenes like cyclohexene and 1-pentenol with $Ag^{+}/S_2O_8^{2-}$ gives hydroxylated cyclohexyl radical **40** and radicals **41**, respectively.[40]

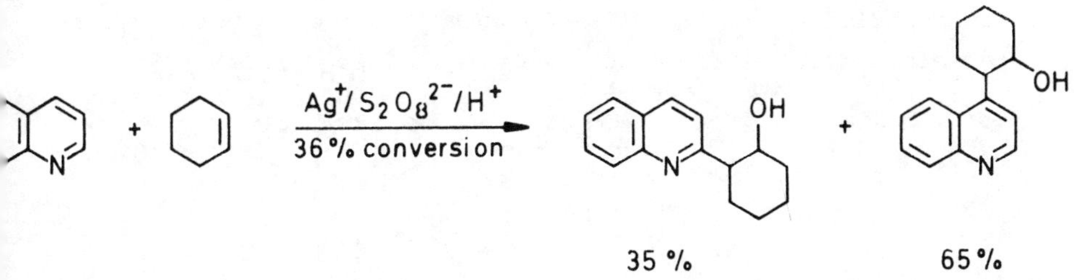
Ag+/S2O8 2-/H+
36% conversion
OH
OH
35 %
65%

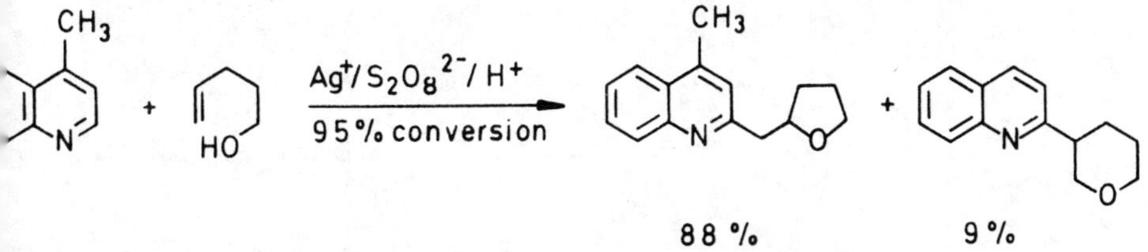
CH3
HO
Ag+/S2O8 2-/ H+
95% conversion
CH3
88 %
9 %

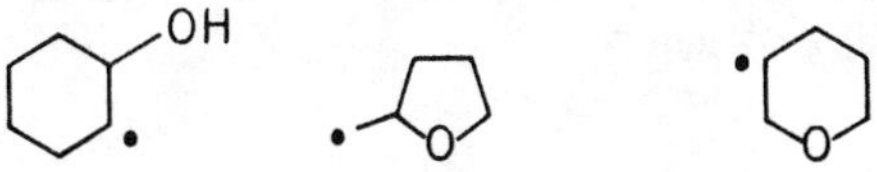
OH

40 **41a** **41b**

e. Derivatives of aldehydes and ketones

Cyclic acetals of formaldehyde like trioxane can be used to introduce dialkoxyalkyl groups, via radical **42**, into heterocycles.[6]

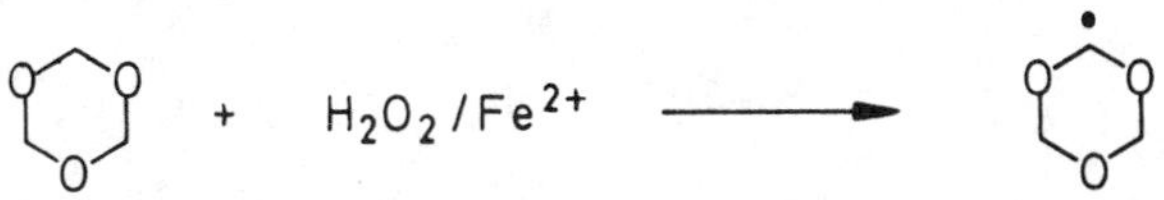
H2O2/Fe2+

42

Ketone derivatives **43** and **45** undergo ring cleavage under oxidative conditions to give substitution products **44** and **46**, respectively.[41]

43 + (cyclohexane with CH_3O and OOH) $\xrightarrow[H_2SO_4]{Fe^{2+}}$ 44 [$(CH_2)_5CO_2CH_3$ on acridine]   67%

quinoline + **45** (cyclohexane with O—N–CH_3) $\xrightarrow[H_2SO_4]{Fe^{2+}}$ **46** [quinoline–$(CH_2)_5CONHCH_3$] 80

46   (2 : 4 = 53 : 47)

The ring cleavage occurs as a result of ß-fragmentation of radicals **47** and **48**.[41]

43 $\xrightarrow[H_2SO_4]{Fe^{2+}}$ **47** (cyclohexane with CH_3O and $O^{\bullet}$) $\longrightarrow$ $\overset{\bullet}{C}H_2\text{-}(CH_2)_4\text{-}CO_2CH_3$

45 $\xrightarrow[H_2SO_4]{Fe^{2+}}$ **48** (cyclohexane with HO and $\overset{\bullet}{N}$–CH_3) $\longrightarrow$ $\overset{\bullet}{C}H_2\text{-}(CH_2)_4\text{-}CONHCH_3$

f. Carboxylic acids

A large number of prim., sec., and tert. alkyl radicals can be
obtained from the corresponding carboxylic acids by reaction
with $Ag^+/S_2O_8^{2-}$.[6,42]

$$S_2O_8^{2-} \; + \; Ag^+ \; \longrightarrow \; SO_4^{-\bullet} \; + \; SO_4^{2-} \; + \; Ag^{2+}$$

$$SO_4^{-\bullet} \; + \; Ag^+ \; \longrightarrow \; SO_4^{2-} \; + \; Ag^{2+}$$

$$RCO_2H \; + \; Ag^{2+} \; \longrightarrow \; RCO_2^{\bullet} \; + \; H^+ \; + \; Ag^+$$

$$RCO_2^{\bullet} \; \longrightarrow \; R^{\bullet} \; + \; CO_2$$

Substitution reactions with these alkyl radicals are very
useful in heterocyclic chemistry.[6,43,44]

$$\text{imidazole} + (CH_3)_3CCO_2H \xrightarrow[S_2O_8^{2-}]{Ag^+/H^+} \text{2-}C(CH_3)_3\text{-imidazole}$$

80%

$$\text{pyridazine} + \text{HO}_2C\text{-pyrrolidine-Ac} \xrightarrow[S_2O_8^{2-}]{Ag^+} \text{product}$$

57%

Arylperoxides, alkylpercarbonates, and perborates can be used in place of peroxydisulfate to oxidize Ag^+ to Ag^{2+}. [45]

g. Dialkylamides

Dialkylformamides are oxidized with $Fe^{2+}/S_2O_8^{2-}$ predominately to amidoalkyl radicals via electron transfer reactions. [46]

$$S_2O_8^{2-} + Fe^{2+} \longrightarrow SO_4^{2-} + SO_4^{\overline{\bullet}} + Fe^{3+}$$

$$SO_4^{\overline{\bullet}} + H\text{-}\underset{O}{\overset{\|}{C}}\text{-}N(CH_3)_2 \longrightarrow SO_4^{2-} + H\text{-}\underset{O}{\overset{\|}{C}}\text{-}\overset{+}{N}(CH_3)_2$$

$$H\text{-}\underset{O}{\overset{\|}{C}}\text{-}\overset{+}{\underset{\bullet}{N}}(CH_3)_2 \longrightarrow H\text{-}\underset{O}{\overset{\|}{C}}\text{-}N\overset{\overset{\bullet}{C}H_2}{\underset{CH_3}{\diagdown}} + H^+$$

$$\text{4-methylquinoline} + H\text{-}\overset{O}{\overset{\|}{C}}\text{-}N(CH_3)_2 \xrightarrow[S_2O_8^{2-}]{Fe^{2+}/H^+} \text{product}$$

98%

Alkoxy radicals abstract a hydrogen atom from the carbonyl
carbon atom of formic amides, but N-alkylamides from other
carboxylic acids give rise to α-N-amidoalkylation.[47]

2. Acylation

a. Formation of aldehydes

The introduction of a formyl group into a heteroaromatic
compound can best be carried out with trioxane or 1,3-dioxolane,
which after undergoing the C-C bond formation give formylated
heterocycles.[48]

b. Formation of ketones

Heterocyclic ketones are synthesized via hydrogen abstraction
from aldehydes[49] or oxidative decarboxylation of α-ketoacids.[50]

c. Formation of carboxylic acids

Formic acid derivatives, keto ester or oxalic ester precursors
are used to introduce carboxylic acid derivatives into
heteroaromatic compounds.[51]

Various syntheses have been carried out using these methods. For example, Rüchardt[52] synthesized fusarinic acid **50** from pyridine derivative **49**.

$$\text{82\%}$$

49   **50**

3. Miscellaneous C-C bond formations

Although alkylation and acylation reactions of Minisci and his group are the most important radical C-C bond forming reactions with heterocycles, some other methods have also been used. Thus, purines can be methylated with tert-butyl peracetate.[53]

$$\text{83\%}$$

Photolysis of isopropylalcohol in the presence of 6-ethoxy-purine gives a mixture of mono- and disubstituted products.[54]

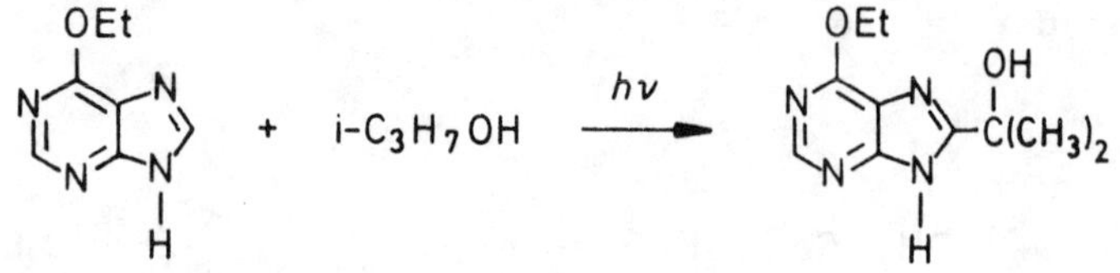

41 %

Peroxide initiated reactions of $BrCCl_3$ introduced perchloroalkyl substituents into purines.[55]

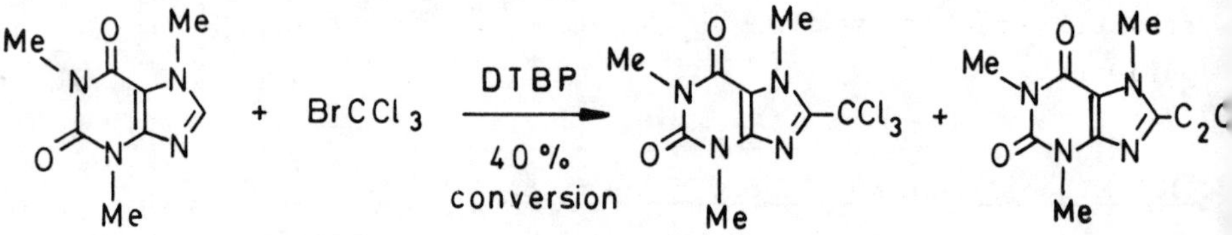

62 %   28 %

In an intramolecular reaction stannane reduction of the adenosine derivative **51** leads to product **52**.[56]

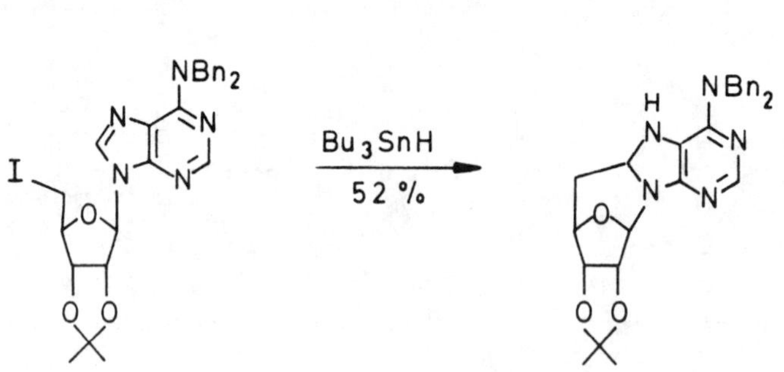

51   52

Radicals generated from acids by the method of Barton[57]
(see p. 82)   also lead to substitution reactions with hetero-
aromatics.

In a photostimulated reaction, Russell[58] produced alkyl radicals
from alkylmercury salts (see p. 96) that add to unprotonated
pyridine.

Heterocyclic aromatic compounds also undergo radical ipso
substitution reactions, especially with σ-radicals and
heterocycles that are protonated or substituted with electron
withdrawing substituents.[28,59] Efficient leaving groups are
RCO, NO_2, and RSO_2.

$$X = NO_2, \qquad R = Ad \qquad (95\%)$$
$$X = COEt, \qquad R = Ad \qquad (60\%)$$
$$X = COEt, \qquad R = MeCO \quad (73\%)$$
$$X = SO_2C_6H_5, \ R = Ad \qquad (80\%)$$

The amount of ipso substitution is always lower with π-radicals than with σ-radicals. According to Tiecco,[60] the substituted thiophene **53** gives ipso substitution product **54** with adamantyl radicals, but product **55** with methyl radicals.

	53	**54**	**55**
R =	Ad :	formed	not formed
R =	CH$_3$:	not formed	formed

Ipso substitution can also lead to rearranged products if 1,2-shifts of ipso intermediates occur. Thus, thiophene **56** reacts with adamantyl radicals to give **57** and **58**. The inter- mediate for both products is presumably radical **59**, which either loses the formyl group or undergoes rearrangement.[61]

56   57   58

59

D. Reactions of Aryl Radicals

Although aryl radicals can be generated in various reactions, three methods have turned out to be most useful for synthetic C-C bond forming purposes:

- Reactions of diazonium salts with cuprous chloride in the presence of alkenes (Meerwein reaction).

- Reaction of halides with stannane in the presence of alkenes.

- Reaction of halides, phosphonates, sulfides, or ammonium salts with carbanions under electron transfer conditions ($S_{RN}1$ reaction).

$$ArI \quad + \quad \diagdown = \diagup_Y \quad \xrightarrow{Bu_3SnH} \quad Ar\diagdown\diagup\diagdown_Y$$

$$Ar X \quad + \quad {}^-C{\diagdown}^{\diagup} \quad \xrightarrow{e^-} \quad Ar{-}C{\diagdown}^{\diagup}$$

1. Aryldiazonium salts/metal ions

The reaction of aryldiazonium salts with CuCl produces aryl radicals which are trapped by alkenes to give adduct radicals **60**. These radicals abstract a chlorine atom from $CuCl_2$ to give product **61** and CuCl. Thus, the Meerwein reaction is a chain reaction in which catalytic amounts of Cu(I) and Cu(II) salts are involved as electron and halogen donors.[62]

Carbocyclic aromatic compounds containing electron-withdrawing
or electron-donating substituents have been used as diazonium
salts. In reactions with styrene, the yields of C-C bond
formation reactions depend only slightly on the nature of the
substituent of the phenyldiazonium salts.[63]

$$X\!\!-\!\!\langle \bigcirc \rangle\!\!-\!\!N_2^+Cl^- + H_2C\!\!=\!\!CHC_6H_5 \xrightarrow{CuCl_2} X\!\!-\!\!\langle \bigcirc \rangle\!\!-\!\!CH_2\!\!-\!\!\overset{\overset{\displaystyle Cl}{|}}{C}HC_6H_5$$

$$X = p-NO_2(80\%), \; m-NO_2 \; (56\%),$$
$$o-NO_2(46\%), \; p-Cl \; (68\%),$$
$$p-Br \; (74\%), \quad H \; (44\%),$$
$$p-CH_3(52\%), \; p-OCH_3 \; (42\%)$$

With more polar alkenes like α-chloroacrylonitrile, the same
substituent independence of the diazonium salt is observed.[64]

$$X\!\!-\!\!\langle \bigcirc \rangle\!\!-\!\!N_2^+Cl^- + H_2C\!\!=\!\!CClCN \xrightarrow{CuCl_2} X\!\!-\!\!\langle \bigcirc \rangle\!\!-\!\!CH_2CCl_2CN$$

$$X = p-NO_2(65\%), \; o-NO_2 \; (62\%), \; p-Cl \; (68\%),$$
$$p-Br \; (68\%), \; H \; (53\%),$$
$$p-CH_3(52\%), \; p-CH_3O \; (53\%)$$

Aryl radicals are so reactive that a large variety of alkenes
is suitable for these syntheses. Alkenes substituted with
chlorine, acetoxy, phenyl, pyridyl, aldehyde, ketone,
carboxylic acid, ester, amide, nitrile, sulfone, and phos-
phonate substituents have been used; even ethylene itself
reacts.[62]

$$O_2N-\!\!\left\langle\bigcirc\right\rangle\!\!-N_2{}^+Cl^- \ + \ \overset{Y}{\underset{Z}{=}\!\!<} \ \xrightarrow{\text{CuCl}_2} \ O_2N-\!\!\left\langle\bigcirc\right\rangle\!\!-CH_2CClYZ$$

X=Y=H (50%) ; X=H, Y=Cl (62%) ; X=Y=Cl (70%) ; X=H, Y=OAc (41%

X=H, Y=C$_6$H$_5$ (80%) ; X=H, Y=COCH$_3$ (41%) ; X=H, Y=CO$_2$CH$_3$ (56%) ;

X=H, Y=CONH$_2$ (64%) ; X=CH$_3$, Y=CN (55%); X=H, Y=PO(OEt)$_2$ (17%) ;

X=H, Y=SO$_2$CH$_3$ (48%)

In reactions with 1,2-disubstituted alkenes and alkynes, regio-
isomers and stereoisomers, respectively, are formed. In some
cases, elimination reactions occur during or after the radical
steps.[62] With cyclic alkenes the yields are often low, but
with the more reactive dienes, reasonable yields can be
obtained.[62,65]

In most cases, the diazonium salts are prepared in situ from aromatic amines either in aprotic solvents with alkylnitrites[66] or in acidic aqueous medium with $NaNO_2$.[67] Generally, the yields of the Meerwein reaction are higher in aprotic organic solvents than in mixtures containing water.

$$CH_3-\langle\bigcirc\rangle-NH_2 \;+\; H_2C=CHCN \xrightarrow[\text{or b)}]{\text{a)}} \; CH_3-\langle\bigcirc\rangle-CH_2CHClCN$$

a) $(CH_3)_3CONO/CuCl_2/CH_3CN$ (73%)

b) $NaNO_2/HCl/CuCl_2/H_2O$ (40%)

In an alternative to the Meerwein reaction, Citterio[68] used Ti^{3+} salts as reagents. The Ti^{3+} salts reduce diazonium salts to aryl radicals, and adduct radicals **60** to anions **62** which are protonated to give product **63**. Thus, Ti^{3+} is not a catalyst and has to be used in excess.

$$\langle\bigcirc\rangle-N_2^+ \;+\; \text{(methyl vinyl ketone)} \xrightarrow[75\%]{Ti^{3+}} \text{(4-phenyl-2-butanone)}$$

$$\langle\bigcirc\rangle-N_2^+ \;+\; NC\diagdown\diagup CN \xrightarrow[69\%]{Ti^{3+}} \text{(aryl succinonitrile)}$$

$$ArN_2^+ \xrightarrow{Ti^{3+}} Ar\bullet \;\; \diagup\diagdown_Y \;\; Ar\diagdown\diagup\diagdown_Y \xrightarrow{Ti^{3+}} Ar\diagdown\diagup\diagdown_Y^- \xrightarrow{H^+} Ar\diagdown\diagup\diagdown_Y$$

60   **62**   **63**

2. Arylhalide / tin hydride

The radical chain reaction using Bu_3SnH (see p. 56) has also been established with aryl halides.[69] Trapping of aryl radicals with electron-poor alkenes leads to adduct radicals **60** which react with stannane to give products **63**.

The yields of these reactions are often not very high.[69]

ArI : C_6H_5I (46%), $p-NH_2C_6H_4I$ (50%), (48%)

(45%)

Bromides can be used in cyclization reactions. Thus, the reaction of bromide **64** with Bu_3SnH gives **65**. The product forming step is the ß-bond cleavage of radical **67**.[70]

$$64 \xrightarrow[\text{AIBN}/\Delta]{\text{Bu}_3\text{SnH}} 65$$

X = NH (96 %) , X = O (75 %)

$$64 \xrightarrow{\text{Bu}_3\text{Sn}\bullet} 66 \longrightarrow 67 \xrightarrow[-\text{C}_6\text{H}_5\text{S}\bullet]{} 65$$

3. $S_{RN}1$ reaction

In 1970 Bunnett[71] observed that the amination of 5- and 6-iodo-pseudocumene in liquid amonia with catalytic amounts of potassium yields almost exclusively substitution   products **68** and **69,** respectively.

$$\xrightarrow[\text{50 \%}]{\begin{array}{c}\text{KNH}_2\\\hline \text{K}/\text{NH}_3\end{array}} 68$$

69

This very high regioselectivity, which is not in accord with
an aryne intermediate, can be explained by a chain reaction
in which aryl radicals attack nucleophiles Y^- to give radical
anions **70**. These radical anions transfer an electron to the
educts and yield substitution products **71**. Cleavage of the
C-X bond of the newly formed radical anions **72** regenerate
the aryl radicals.[71]

This $S_{RN}1$ scheme is analogous to that proposed by Kornblum[72]
and Russell[73] for reactions of nucleophiles with certain
p-nitrobenzyl halides and 2-halo-2-nitropropanes. Reactions

with carbanions offer a wide variety of C-C bond formation possibilities with carbo- and heteroaromatic systems.

Halogens, phosphates, phenylsulfides, and trimethylammonium groups have been used as nucleofugic groups which can be substituted in these syntheses. These reactions also occur in the presence of substituents like alkyl, alkoxy, phenyl, carboxylate, and benzoyl groups, but nitrile and hydroxy groups are incompatible.[74]

A variety of carbo- and heteroaromatic substrates, equipped with suitable nucleofugic groups, participate in $S_{RN}1$ reactions. These include benzenes, naphthalenes, anthracenes, phenanthrenes, pyridines, pyrimidines, pyrazines, quinolines, iso-quinolines, and thiophenes.[74]

Liquid ammonia has turned out to be the most suitable solvent, but dimethyl sulfoxide, tetrahydrofuran , and aqueous tert-butyl alcohol have also been employed with some success. The syntheses are promoted by electron donation by potassium metal or, in few cases, from electrodes, and irradiation to stimulate the electron transfer steps.[74]

The carbanions used as nucleophiles in these C-C bond forming reactions shall be discussed in detail.

a. Conjugated hydrocarbons

Conjugated hydrocarbons that have lower pK_a values than ammonia (pK_a = 32.5) are converted into their carbanions by reaction with amide ions in liquid ammonia. Many of these carbanions can be used for C-C bond formation reactions if they are soluble in liquid ammonia. Thus, 1,3-pentadiene reacts with phenylbromide to give, after hydrogenation, 1-phenylpentane;[75] picolines can be arylated at the methyl group.[76]

1) $NH_2^-/NH_3/K$

2) H_2 / Pt

74 %.

The reaction with 2-bromomesitylene shows that $S_{RN}1$ reactions are not hindered by small ortho substituents.

b.   Ketones

The best procedure for conducting arylation reactions of ketones involves the generation of carbanions with potassium amide or potassium tert-butoxide in liquid ammonia, followed by addition of the aromatic substrate and photolysis.[74,77]

Heterocyclic halides give comparable yields of C–C bond formation products.[78]

Cyclization reactions can also be carried out using the $S_{RN}1$ method. Thus, ($\pm$) cephalotoxinone **74** has been synthesized from iodide **73**.[79]

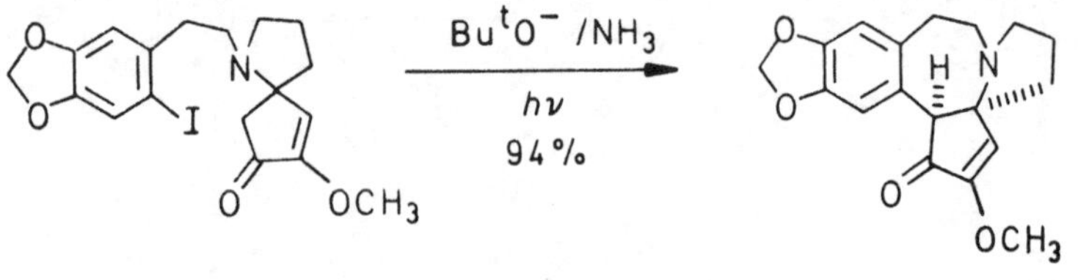

73   74

Medium sized rings can also be synthesized in intramolecular
C-C bond forming reactions.[80]

n = 1 (99%)

n = 3 (73%)

n = 5 (35%)

Ortho-substituted aromatic compounds give rise to cyclic
products by successive reactions.[81]

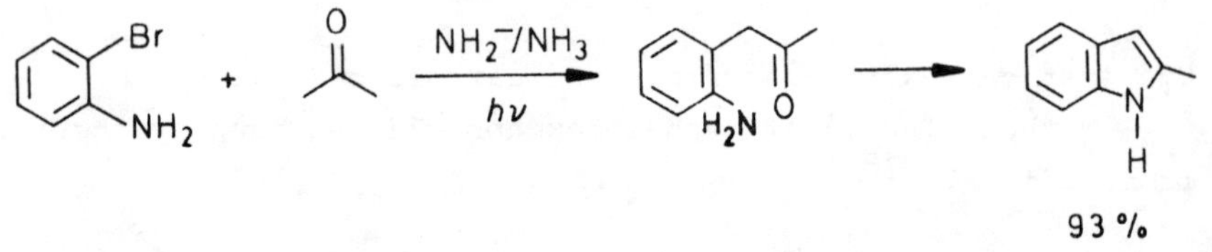

93%

40 %

In a reaction involving o-dibromobenzene, the product of two $S_{RN}1$ reactions cyclizes to isomeric indenes.[82]

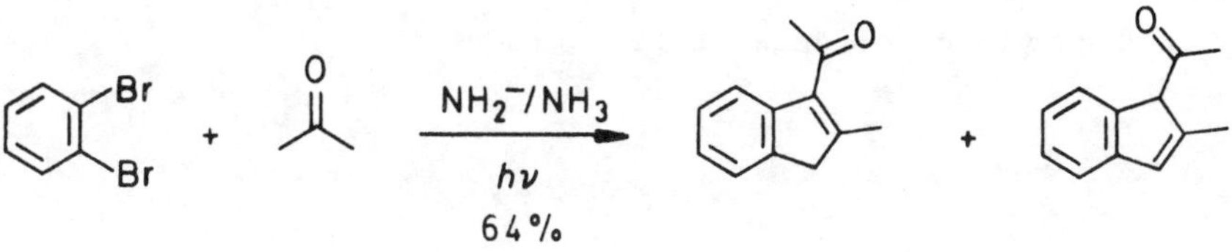

64 %

c. Aldehydes, esters, and amides

Aldehydes and esters are not as useful in $S_{RN}1$ syntheses as ketones. Reduction of the halobenzene with aldehydes and disubstitution with esters severely limits these reactions.[80,83]

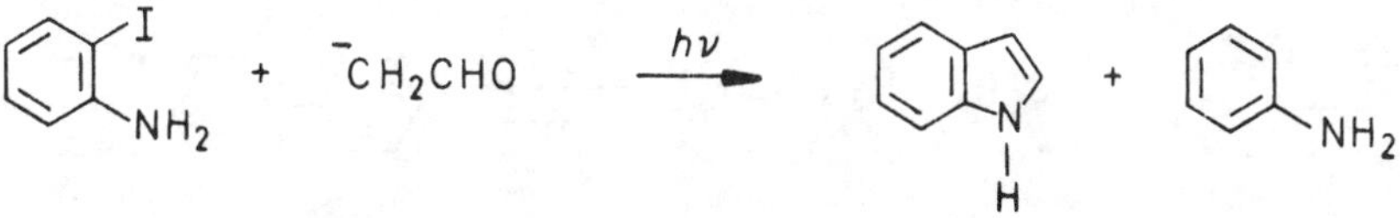

$$MeO{-}\langle\bigcirc\rangle{-}Br \quad + \quad {}^{-}CH_2CO_2Bu^t \quad \xrightarrow{h\nu}$$

$$MeO{-}\langle\bigcirc\rangle{-}CH_2{-}CO_2Bu^t \quad 67\%$$

$$+$$

$$\left(MeO{-}\langle\bigcirc\rangle{-}CHCO_2Bu^t\right)_2 \quad 29\%$$

The acetamide anion fails to react with arylhalides in liquid ammonia; however, the enolate ions of N,N-disubstituted amides give high yields of substitution products.[84]

$$\langle\bigcirc\rangle{-}Br \quad + \quad (CH_3)_2NCCH_2^{-} \quad \xrightarrow[\substack{h\nu \\ 72\%}]{NH_3} \quad \langle\bigcirc\rangle{-}CH_2CN(CH_3)_2$$
(with $\overset{O}{\overset{\|}{}}$ above each carbonyl)

$$+ \quad (C_6H_5)CH_3NCCH_2^{-} \quad \xrightarrow[\substack{h\nu \\ 80\%}]{NH_3}$$

Intramolecular C-C bond formation reactions are also possible
with N,N-disubstituted amides. These reactions have been used
for the synthesis of oxindole **76** from **75** in a tetrahydrofuran/
hexane mixture, with lithium diisopropyl amide as base.[85]

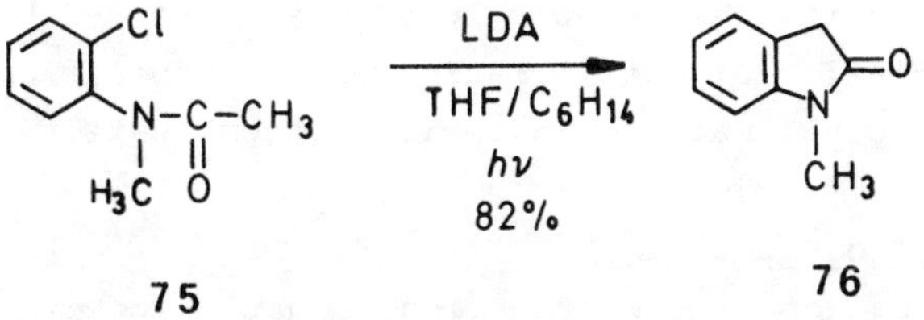

75   **76**

E. Radical Ions

Electron transfer to or from benzoid systems leads to radical
ions **77** and **78** with 5 and 7 π-electrons, respectively.

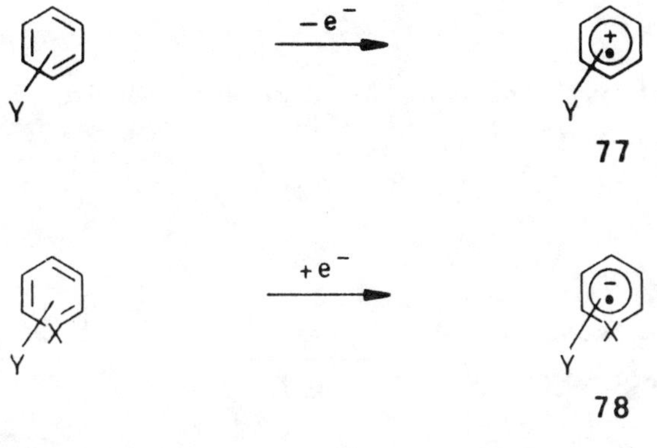

77

78

Radical cations **77** are intermediates in most side chain
oxidations carried out at the anode or with metal salts.[86]
Some electrophilic aromatic substitution reactions are also
believed to proceed via these cations.[87] The formation of
C-C bonds occurs via recombination or reaction of **77** with
carbon nucleophiles.

Radical anions **78** are intermediates of $S_{RN}1$ reactions which
lead, with suitable nucleofugic groups Y, to C-C bonds via
phenyl radicals. Direct formation of C-C bonds from radical
anions **78** occurs mainly via radical recombination.

1. Radical cations

Anodic oxidation of electron-rich aromatic systems leads to
aromatic radical cations which recombine to give biphenyls
after deprotonation.[88]

Thus, 9-methoxyanthracene and N-methylcarbazole can be
oxidatively coupled.[89]

With phenols, radical cations deprotonate to give phenoxy
radicals which combine to give dimers.[88]

Thus, tetrahydroisoquinoline **79** was coupled to give the
dimer **80**.[90]

Intramolecular C-C bond formation reactions between aromatic rings occur by trapping the radical cations with suitably situated aromatic compounds. The newly formed radical is then further oxidized; if aromatization is not possible, quinoid systems are formed.[91]

This method has been used in the synthesis of morphinandienone **82** from tetrahydroisoquinoline **81**.[92]

Aromatic radical cations can also be trapped intermolecularly by anions; thus, nitriles are formed in the presence of cyanide ions.[93]

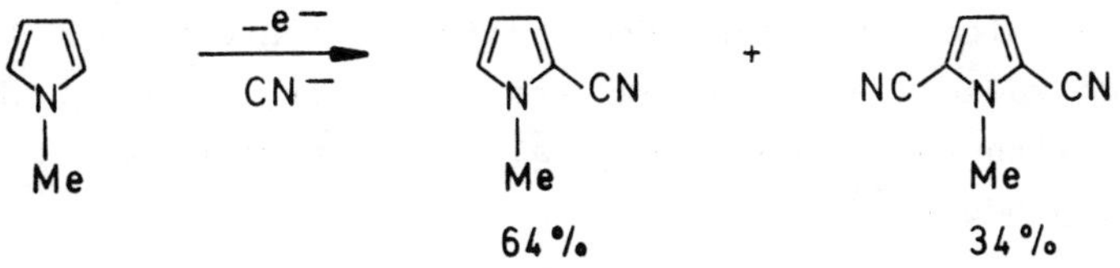

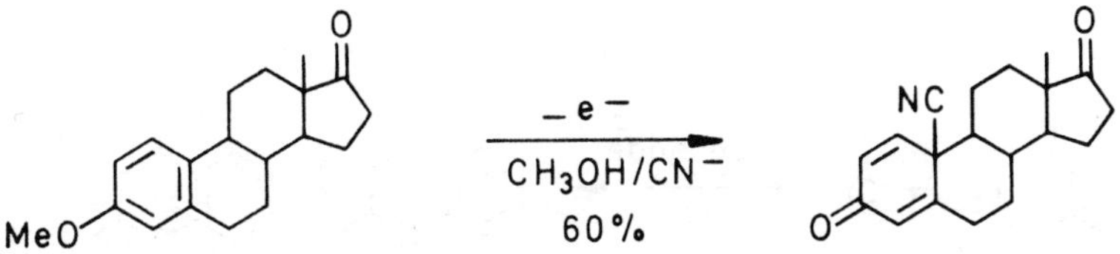

64 %   34 %

60 %

2. Radical anions

Direct C-C bond forming reactions from radical anions are scarce. They occur via radical combination reactions in a complex reaction scheme.

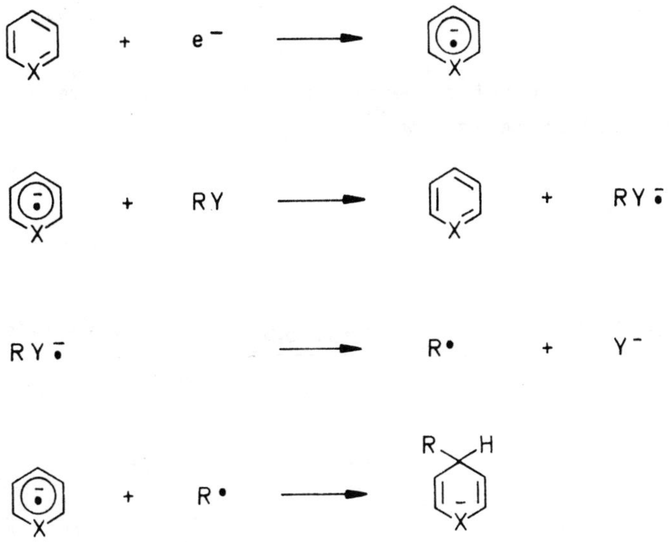

The radical anion, formed at the cathode or with a metal, transfers an electron to an alkyl halide. The newly formed alkylhalide radical anion decomposes to an alkyl radical which recombines with the aromatic radical anion; protonation then yields the product.[94]

Tertiary alkyl halides are mainly used and the radical combination often occurs with a low selectivity. Thus, α-methyl-quinoline leads to a mixture of products.[95]

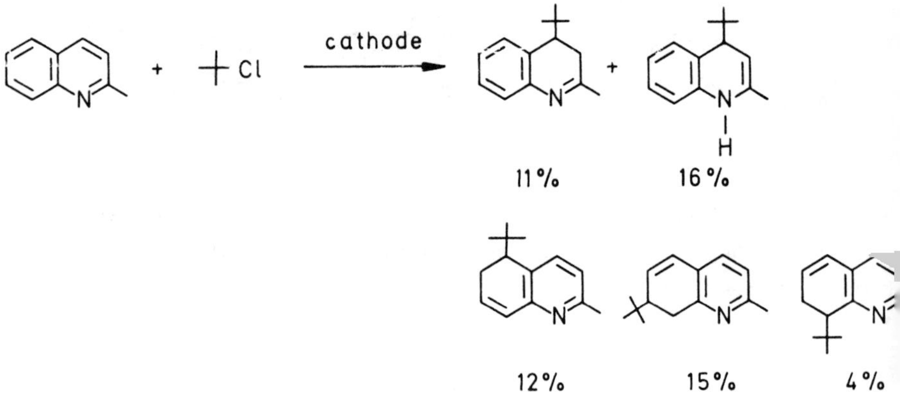

11%   16%

12%   15%   4%

The reaction of nicotinic amide 83 is more selective, but the yield of product 84 is low.[96]

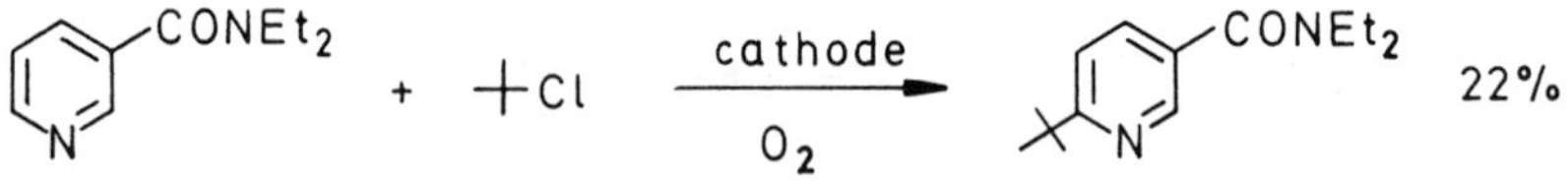

22%

83   84

REFERENCES

1. A. Citterio, F. Minisci, O. Porta, G. Sesana, *J. Am. Chem. Soc.* **1977**, *99*, 7960.

2. K. Münger, H. Fischer, *Int. J. Chem. Kinetics* **1985**, *17*, 809.

3. B. Giese, *Angew. Chem. Int. Ed. Engl.* **1983**, *22*, 771.

4. T.J. Burkey, D. Griller, L. Lunazzi, A.S. Nazran, *J. Org. Chem.* **1983**, *48*, 3704.

5. F. Minisci, A. Citterio in G.H. Williams (ed.): *Advances in Free-Radical Chemistry* Vol. VI, Heyden, London 1980, p. 65.

6. F. Minisci, *Synthesis* **1973**, 1; F. Minisci, *Top. Curr. Chem.* **1976**, *62*, 1.

7. M.E. Kurz, P. Kovacic, A.K. Bose, J. Kugajevsky, *J. Am. Chem. Soc.* **1968**, *90*, 1818; M.E. Kurz, E.M. Steele, R.L. Vecchio, *J. Org. Chem.* **1974**, *39*, 3331.

8. M.E. Kurz, V. Baru, P.N. Nguyen, *J. Org. Chem.* **1984**, *49*, 1603.

9. M.E. Kurz, P. Ngoviwatchai, T. Tantrarant, *J. Org. Chem.* **1981**, *46*, 4668.

10. E. Bamberger, *Ber. Dtsch. Chem. Ges.* **1897**, *30* 366.

11. W.S.M. Grieve, D.H. Hey, *J. Chem. Soc.* **1934**, 1797.

12. D.H. Hey, C.J.M. Stirling, G.H. Williams, *J. Chem. Soc.* **1956**, 1475.

13. R. Huisgen, G. Horeld, *Liebigs Ann. Chem.* **1949**, *562*, 137; R. Huisgen, H. Nakaten, *Liebigs Ann. Chem.* **1951**, *573*, 181.

14. C. Rüchardt, B. Freudenberg, *Tetrahedron Lett.* **1964**, 3623; G. Binsch, E. Merz, C. Rüchardt, *Chem. Ber.* **1967**, *100*, 247.

15. M.J. Perkins in J.K. Kochi (ed.): *Free Radicals*, Wiley, New York 1973, p. 231.

16. J.I.G. Cadogan, *Acc. Chem. Res.* **1971**, *4*, 186.

17. J.I.G. Cadogan in G.H. Williams (ed.): *Advances in Free Radical Chemistry*, Heyden, London 1980, p. 185.

18. J.I.G. Cadogan, *Pure Appl. Chem.* **1967**, *15*, 153

19. D.H. Hey, M.J. Perkins, G.H. Williams, *J. Chem. Soc.* **1965**, 110.

20. C.S. Rondestvedt, H.S. Blanchard, *J. Am. Chem. Soc.* **1955**, *77*, 1769.

21. M.J. Perkins in J.K. Kochi (ed.): *Free Radicals* Vol. II, Wiley,New York 1973, p. 231.

22. P.J.Bunyan, D.H. Hey, *J. Chem. Soc.* **1960**, 3787.

23. M. Hasebe, K. Kogawa, T. Tsuchiya, *Tetrahedron Lett.* **1984**, *25*, 3887.

24. R.K. Sharma, N. Kharasch, *Angew. Chem. Int. Ed. Engl.* **1968**, *7*, 36; I. Grimshaw, A.P. de Silva, *Chem. Soc. Rev.* **1981**, *10*, 181.

25. R. Ito, T. Migita, N. Morikawa, O. Simamura, *Tetrahedron* **1965**, *21*, 955.

26. B. Chauncy, E. Gellert, *Aust. J. Chem.* **1970**, *23*, 2503.

27. M. Sainsbury, *Tetrahedron* **1980**, *36*, 3327.

28. M. Tiecco, *Acc. Chem. Res.* **1980**, *13*,51.

29. C.M. Camaggi, R. Leardini, P. Zanirato, *J. Org. Chem.* **1977**, *42*, 1570.

30. J.J. Köhler, W.N. Speckamp, *Tetrahedron Lett.* **1977**, 631.

31. A. Citterio, F. Minisci, V. Franchi, *J. Org. Chem.* **1980**, *45*, 4752.

32. M. Bellati, T. Caronna, A. Citterio, F. Minisci, *J. Chem. Soc. Perkin Trans. 2* **1976**, 1835.

33. C. Giordano, F. Minisci, V.Tortelli, E.Vismara, *J.Chem.Soc Perkin Trans. 2* **1984**, 293; F. Minisci, C. Giordano, E. Vismara, S. Levi, V. Tortelli, *J. Am. Chem. Soc.* **1984**, *106*, 7146.

34. T. Caronna, A. Citterio, T. Crolla, F. Minisci,
     Tetrahedron Lett. **1976**, 1731.

35. T. Caronna, A. Citterio, T. Crolla, F. Minisci,
     J. Chem. Soc. Perkin Trans.1 **1977**, 865.

36. G. Castaldi, F. Minisci, V.Tortelli,E.Vismara,*Tetrahedron Lett.* **1984**, *25*, 3897.

37. F. Minisci, A. Citterio, E. Vismara, C. Giordano,
     Tetrahedron **1985**, *41*, 4157.

38. C. Walling, *Acc. Chem. Res.* **1975**, *8*, 125.

39. A. Citterio, A. Gentile, F. Minisci, *Tetrahedron Lett.*
     1982, *23*, 5587.

40. A. Clerici, F. Minisci, K. Ogawa, J.M. Surzur,
     Tetrahedron Lett. **1978**, 1149.

41. F. Minisci, R. Galli, V. Malatesta, T. Caronna,
     Tetrahedron **1970**, *26*, 4083.

42. F. Minisci, A. Citterio, C. Giordano, *Acc. Chem. Res.*
     1983, *16*, 27.

43. F. Minisci, R. Bernardi, F. Bertini, R. Galli, M. Perchi-
     nummo, *Tetrahedron* **1971**, *27*, 3575.

44. G. Heinisch, A. Jentzsch, I. Kirchner, *Tetrahedron Lett.*
     1978, 619.

45. F. Minisci, E. Vismara, U. Romano, *Tetrahedron Lett.*
     1985, *26*, 4803

46. G.P. Gordini, F. Minisci, G. Palla, A. Arnone, R. Galli,
     Tetrahedron Lett. **1971**, 59.

47. A. Arnono, M. Cecere, R. Galli, F. Minisci, M. Perchinummo,
     O. Porta, G.P. Gardini, *Gazz. Chim. Ital.* **1973**, *103*, 13.

48. G.P. Gardini, *Tetrahedron Lett.* **1972**, 4113

49. G.P. Gardini, F. Minisci, *J. Chem. Soc. C* **1970**, 929;
     M. Bellati, T. Caronna, A. Citterio, F. Minisci,
     J. Chem. Soc. Perkin Trans. 2 **1976**, 1835.

50. T. Caronna, G. Fronza, F. Minisci, O. Porta, *J. Chem. Soc. Perkin Trans.* 2 **1972**, 2035; T. Sakamoto, S. Konno, H. Yamanaka, *Heterocycles* **1977**, *6*, 1616.

51. F. Minisci, G.P. Gardini, R. Galli, F. Bertini, *Tetrahedron Lett.* **1970**, 15; R. Bernardi, T. Caronna, R. Galli, F. Minisci, M. Perchinunno, *Tetrahedron Lett.* **1973**, 645; A. Citterio, A. Gentile, F. Minisci, M. Serravalle, S. Ventura, *J. Org. Chem.* **1984**, *49*, 3364.

52. E. Langhals, H. Langhals, C. Rüchardt, *Liebigs Ann. Chem.* **1982**, 930.

53. M.F. Zady, J.L. Wong, *J. Org. Chem.* **1979**, 44, 1450.

54. H. Steinmaus, I. Rosenthal, D. Elad, *J. Org. Chem.* **1971**, *36*, 3594.

55. J. Zylber, N. Zylber, A. Chiaroni, C. Riche, *Tetrahedron Lett.* **1984**, *25*, 3853.

56. K.N.V. Duong, A. Gaudemer, M.D. Johnson, R. Quillivic, J. Zylber, *Tetrahedron Lett.* **1975**, 2997.

57. D.H.R. Barton, B. Garcia, H. Togo, S.Z. Zard, *Tetrahedron Lett.* **1986**, *27*, 1327.

58. G.A. Russell, D. Guo, R.K. Khanna, *J. Org. Chem.* **1985**, *50*, 3423.

59. M. Fiorentino, L. Testaferri, M. Tiecco, L. Troisi, *J. Chem. Soc. Chem. Commun.* **1977**, 316; M. Fiorentino, L. Testaferri, M. Tiecco, L. Troisi, *J. Chem. Soc. Perkin Trans.* 2 **1977**, 1679; T. Caronna, A. Citterio, M. Belatti, *J. Chem. Soc. Chem. Commun.* **1976**, 987.

60. L. Testaferri, M. Tiecco, M. Tingoli, M. Fiorentino, L. Troisi, *J. Chem. Soc. Chem. Commun.* **1978**, 93.

61. P. Cogolli, L. Testaferri, M. Tiecco, M. Tingoli, *J. Chem. Soc. Perkin Trans.* 2 **1980**, 1336.

62. C.S. Rondestvedt, *Org. Reactions* **1976**, *24*, 225.

63. K.G. Tashchuk, A.V. Dombrovskii, *J. Org. Chem. USSR* **1965**, 1, 2034.

64. N.O. Pastushak, A.V. Dombrovskii, L.I. Rogovik, *J. Gen. Chem. USSR* **1964**, *34*, 2254.

65. M. Allard, J. Levisalles, *Bull. Soc. Chim. France* **1972**, 1926.

66. M.P. Doyle, B. Siegfried, R.C. Elliott, J.F. Dellaria, *J. Org. Chem.* **1977**, *42*, 2431.

67. C.F. Koelsch, *J. Am. Chem. Soc.* **1943**, *65*, 57.

68. A. Citterio, *Org. Synthesis* **1984**, *62*, 67; A. Citterio, E. Vismara, *Synthesis* **1980**, 291.

69. B. Giese, J.A. Gonzalez-Gomez, unpublished results.

70. Y. Ueno, K. Chino, M. Okawara, *Tetrahedron Lett.* **1982**, *25*, 2575.

71. J.K. Kim, J.F. Bunnett, *J. Am. Chem. Soc.* **1970**, *92*, 7463.

72. N. Kornblum, R.E. Michel, R.C. Kerber, *J. Am. Chem. Soc.* **1966**, *88*, 5662.

73. G.A. Russell, W.C. Danen, *J. Am. Chem. Soc.* **1966**, 88, 5663.

74. R.A. Rossi, R.H. de Rossi: *Aromatic Substitution by the $S_{RN}1$ Mechanism*, ACS Monograph 178, Washington 1983; J.F. Wolfe, D.R. Carver, *Org. Prep. Proc.* **1978**, *10*, 225.

75. R.A. Rossi, J.F. Bunnett, *J. Org. Chem* **1973**, *38*, 3020.

76. J.F. Bunnett, B.F. Gloor, *J. Org. Chem.* **1974**, *39*, 382.

77. R.A. Rossi, J.F. Bunnett, *J. Am. Chem. Soc.* **1972**, *94*, 683; R.A. Rossi, R.H. de Rossi, A.F. Lopez, *J. Am. Chem. Soc.* **1976**, *98*, 1252; J.F. Bunnett, J.E. Sandberg, *J. Org. Chem.* **1976**, *41*, 1702.

78, J.F. Bunnett, B.F. Gloor, *Heterocycles* **1976**, *5*, 377; A.P. Komin, J.F. Wolfe, *J. Org. Chem.* **1977**, *42*, 2481; E.A. Oostveen, H.C. van der Plas, *Rec. Trav. Chim. Pays-Bas* **1979**, *98*, 441.

79. S.M. Weinreb, M.F. Semmelhack, *Acc. Chem. Res.* **1975**, *8*, 158.

80. M.F. Semmelhack, T.M. Bargar, *J. Org. Chem.* **1977**, *42*, 1481.

81. R.R. Bard, J.F. Bunnett, *J. Org. Chem.* **1980**, *45*, 1546;
    R. Beugelmans, H.J. Ginsburg, *J. Chem. Soc. Chem. Commun.*
    1980, 508.

82. J.F. Bunnett, P. Singh, *J. Org. Chem. Ref.* 74, p. 113.

83. R. Beugelmans, G. Roussi, *Tetrahedron* **1981**, *37*, 393.

84. R.A. Rossi, R.A. Alonso, *J. Org. Chem.* **1980**, *45*, 1239.

85. J.F. Wolfe, M.C. Sleevi, R.R. Goehring, *J. Am. Chem. Soc.*
    1980, *102*, 3646.

86. C.J. Schlesener, C. Amatore, J.K. Kochi, *J. Am. Chem. Soc.*
    1984, *106*, 3567; C. Walling, G.M. El-Taliami,K. Amarnath,
    J. Am. Chem. Soc. **1984**, *106*, 7573; E. Baciocchi, T. Del
    Giacco, C. Rol, G.V. Sebastiani, *Tetrahedron Lett.* **1985**,
    26, 541; F. Minsci, A. Citterio, *Acc. Chem. Res.* **1983**,
    16, 27.

87. F. Effenberger, *Chem. in unserer Zeit* **1979**, *13*, 87.

88. S. Torii: *Electroorganic Syntheses, Part I, Oxidations*,
    VCH, Weinheim 1985.

89. S.N. Frank, A. Bard, A. Ledwith, *J. Electrochem. Soc.*
    1968, *90*, 4645; O. Hammerich, V.D. Parker, *Acta Chem.
    Scand.* **1982**, *B36*, 519.

90. J.M. Bobbitt, H. Yagi, S. Shibuya, J.T. Stock, *J. Org.
    Chem.* **1971**, *36*, 3006.

91. E. Kotani, N. Takeuchi, S. Tobinaga, *J. Chem. Soc. Chem.
    Commun.* **1973**, 550.

92. L.L. Miller, F.R. Stermitz, J.R. Falck, *J. Am. Chem. Soc.*
    1971, *93*, 5941; E. Kotani, S. Tobinaga, *Tetrahedron Lett.*
    1973, 4759.

93. K. Yoshida, *J. Am. Chem. Soc.* **1979**, *101*, 2116; K. Ponsold,
    H. Kasch, *Tetrahedron Lett.* **1979**, 4465.

94. J.F. Garst, *Acc. Chem. Res.* **1971**, *4*, 400; T. Lund,
    H. Lund, *Tetrahedron Lett.* **1986**, *27*, 95.

95. C. Degrand, H. Lund, *Acta Chem. Scand.* **1977**, *B31*, 593.

96. C. Degrand, D. Jacquin, P.L. Compagnon, *J. Chem. Research
    (S)* **1978**, 246.

Chapter 6

Methods of Radical Formation

This chapter gives a concise collection of methods, which
lead to carbon centered radicals via rupture of C-X bonds
and which are used in the formation of C-C bonds.

$$R-X \longrightarrow R\bullet \longrightarrow R-\overset{|}{\underset{|}{C}}-$$

Detailed discussions and examples of these methods are found
in this book on the pages cited in brackets.

A. Carbon-Halogen Bonds

1. Alkylhalides and arylhalides

$$R-Hal \longrightarrow R\cdot$$

a) Bu_3SnH, AIBN or $h\nu$ (p. 7 - 11, 39, 57 - 64, 111, 142, 154 - 169, 220 - 221, 238, 246 - 247); $(Bu_3Sn)_2$, $h\nu$ (p. 102, 176); allyltin (p. 98 - 101);vinyltin (p. 102).

b) Bu_3GeH, AIBN (p. 39, 67 - 68, 158).

c) $XCo(dmgH)_2py$, $NaBH_4$ or cathode (p. 179 - 181); $Co(dmgH)_2py$ (p. 102 - 103, 191).

d) Vitamin B_{12}, Zn or cathode (p. 112 - 113, 181).

e) Na or K (p. 124, 247 - 249); Zn (p. 189).

f) Anions (p. 95 - 97, 247 - 255, 259 - 260).

g) Bz_2O_2, Fe^{3+} (p. 226 - 227).

h) $h\nu$ or AIBN (p. 77 - 82, 173 - 176, 220).

B.   Carbon-Oxygen Bonds

1.   Alcohols

$$R-OH \longrightarrow R\cdot$$

a) CS_2, MeI, Bu_3SnH, AIBN (p. 57, 64 - 65).

b) $(COCl)_2$, N-hydroxypiperidine-2-thione, hν or
      AIBN (p. 85).

c) $ClPO(OEt)_2$; $S_{RN}1$ conditions (p. 249).

2. Aldehydes, ketones, and esters

$$\text{>C=O} \longrightarrow \text{>}\overset{\bullet}{\text{C}}\text{-OR}$$

a) N_2H_4; Hg^{2+}, $NaBH_4$ (p. 54 - 56).

b)   Cathode, H+ (p. 122, 187 - 188).

c)   R-H, hν (p. 123, 194 - 196).

d)   Zn, TMSCl (p. 186); Mg, TMSCl (p. 187).

$$\text{>C=O} \quad \longrightarrow \quad \text{>}\overset{\displaystyle\cdot}{\text{C}}\text{-O}^-$$

a) Na (p. 124, 185 - 187, 192 - 193);
    Mg (p. 122); Ti(III) (p. 123).

b) R$_3$N or HMPT, hν (p. 188).

c) Cathode (p. 187 - 188).

C. CARBON-SULFUR AND CARBON-SELENIUM BONDS

1. Alkylsulfides, arylsulfides, alkylselenides, and acyl-selenides

$$R-SC_6H_5$$

$$\text{or} \qquad \longrightarrow \qquad R\bullet$$

$$R-SeC_6H_5$$

a) Bu$_3$SnH, AIBN or hν (p. 57, 65, 158, 160 - 161, 166 - 167).

b) Anions, hν (p. 249).

D. CARBON-NITROGEN BONDS

1. Amines

$$R-NH_2 \longrightarrow R\bullet$$

a) MeI, anion, hν or Na (p. 249 - 250).

2. Nitro compounds

$$R-NO_2 \longrightarrow R\bullet$$

a) Bu_3SnH, AIBN (p. 57, 65 - 67, 169).

b) Anion, hν (p. 95 - 97).

3. Diazonium salts

$$Ar-N_2^+X^- \longrightarrow Ar\cdot$$

a) Cu^+, Ti^{3+} (p. 217 - 219, 242 - 245).

4. Azo compounds

$$R-N=N-R \longrightarrow R\cdot$$

a) Δ or $h\nu$ (p. 126 - 127, 197 - 202, 217).

E.   Carbon-Carbon Bonds

1. Carboxylic acids

$$R-CO_2H \qquad \longrightarrow \qquad R\cdot$$

a)   N-Hydroxypiperidine-2-thione, $h\nu$ (p. 82 - 86, 176 - 177, 239).

b) M^{n+}, peroxide (p. 233 - 234, 236).

c) SO_2Cl_2, H_2O_2 or RO_2H, Δ (p. 125 - 126, 237).

d)   Anode (p. 113 - 117, 184).

2. Ketones

$$R-\overset{\overset{\displaystyle O}{\displaystyle \|}}{C}- \quad \longrightarrow \quad R\cdot$$

a) H_2O_2, Fe^{2+} (p. 93 - 94, 232, 236 - 237).

3. Cyclopropanes

a) Hg^{2+}, $NaBH_4$ (p. 49 - 54).

4. Alkenes

a) Hg^{2+}, $NaBH_4$ (p. 43 - 49, 151 - 154).

b) B_2H_6, O_2 (p. 86 - 89); B_2H_6, Hg^{2+}, $NaBH_4$ (p. 41 - 43).

c) Peroxide, M^{n+} (p. 21, 230 - 231).

d) Capto dative alkenes (p. 119 - 120).

e) Cyclization of dienes (p. 172 - 175, 177 - 178, 183 - 184; tandem cyclization (p. 26, 155 - 158, 170); combination of intramolecular and intermolecular reactions (p. 111, 169).

5. Alkenes and aromatic compounds

a)   Anode (p. 117 - 118, 255 - 259).

a) Cathode (p. 124, 259 - 260).

F. Carbon-Hydrogen Bonds

$$R-H \quad \longrightarrow \quad R\cdot$$

a) Peroxides (p. 6, 17 - 18, 69 - 76, 106 - 107, 109 - 110, 119 - 120, 170 - 173, 227 - 231, 234 - 237).

b) Ketone/hν or hν (p. 69 - 76, 120 - 121, 123, 172 - 173, 194 - 196).

c) M^{n+} (p. 89 - 93, 182 - 183, 214 - 216).

d) $R_2\overset{+}{N}HCl$, Fe^{2+} (p. 225 - 226).

e) Δ (p. 69).

G. Carbanions

$$R^- \longrightarrow R\bullet$$

a) Anode (p. 94 - 95, 116 - 117).

b) I_2 or O_2 (p. 118).

H. Carbon-Boron Bonds

$$R-B\diagdown \longrightarrow R\bullet$$

a) O_2 (p. 86 - 89).

b) Hg^{2+}, $NaBH_4$ (p. 41 - 43).

I. Carbon-Mercury Bonds

$$R-HgX \longrightarrow R\cdot$$

a) $NaBH_4$ or Bu_3SnH (p. 23, 38 - 56, 151 - 154, 178).

b) Anions, $h\nu$ (p. 98); heteroaromatic, $h\nu$ (p. 239).

J. Carbon-Cobalt Bonds

$$R-Co \longrightarrow R\cdot$$

a) Δ (p. 111 - 113, 179 - 181).

Author Index

Bold figures refer to the pages on which the authors are cited in the text, roman figures refer to pages on which references are listed at the end of each article.

282

Subject Index